The First Space War

The First Space War

How the Patterns of History and the Principles of STEM Will Shape Its Form

J. Furman Daniel III
and T. K. Rogers

LEXINGTON BOOKS
Lanham • Boulder • New York • London

Published by Lexington Books
An imprint of The Rowman & Littlefield Publishing Group, Inc.
4501 Forbes Blvd., Ste. 200, Lanham, MD 20706
www.rowman.com

6 Tinworth Street, London SE11 5AL, United Kingdom

British Library Cataloguing in Publication Information Available

Library of Congress Cataloging-in-Publication Data available

ISBN: 978-1-4985-8774-7 (cloth)
ISBN: 978-1-4985-8775-4 (electronic)

To our families

Contents

Acknowledgments

Over the past three decades, the two of us have been talking about science, history, self-defense, politics, sustainable farming, art, and a wide range of assorted topics ranging from banal to erudite. Along the way, we have pushed each other to think more critically and ask questions about the universe around us. This book is in many ways a reflection of this friendship and our mutual love of learning.

Despite our best efforts, this book would have been impossible without many other friends, teachers, mentors, critics, publishers, editors, anonymous peer reviewers, and family members. While no single list can hope to mention every person who has had a positive impact on our lives, we would like specifically to thank the following people who have had an impact on this work: Dale Avery, Bryce Dickey, Thomas "T-Dawg" Foley, Tyrone Groh, Bob McKenzie, Paul Musgrave, Joseph Parry, Bryndee Ryan, Brooke Shannon, Joanna Spear.

Special thanks are also due to our families, most notably: Mary Berry, Christina Capacci-Daniel, Kelly Rogers Farabee, and Sandy Rogers with an additional recognition of Scott and Mark Rogers's invaluable role in establishing intuitor.com over two decades ago, which among many things has, over the years, reviewed movie depictions of space battles, and for their many discussions about Mars. These activities were important steps on the path to this current work. Our families' love, patience, humor, editing, thinking, and support have made this a true pleasure. We love you all and dedicate this book to you.

Introduction

A WAR WITH MARTIANS?

Using physics, history, and politics as guides, this book provides a detailed account of how Earth's first war in space likely will be fought. Unfortunately, much of what people believe about space warfare has been shaped, or should we say misshaped, by Hollywood and other forms of popular media. In the following book a STEM educator and a political science professor team up to explore the possibilities and explain why almost everything you've learned about space wars from pop culture is disappointingly wrong.

When we first discussed writing about this topic, we thought in terms of popular science fiction films that depicted space warfare in "far, far away" galaxies, or at least in faraway parts of our own galaxy, but we wanted something more down-to-earth, like an invasion. Invasions of nefarious aliens from other star systems, such as portrayed in the movie *Independence Day*, are great fun, but again, are from far, far away, maybe a bit too far away for our purposes. The solution: Martians. What other group of otherworldly beings have gained such a foothold in the collective imagination of humanity?

With H. G. Wells's 1898 book *The War of the Worlds*, Martians were launched into the universe of pop culture like no other extraterrestrials. The book's repeated repackaging in movies—and most notably in Orson Welles's famous radio broadcast on the night before Halloween in 1938—has fueled fascination, even terror, among humans. Martians have become our favorite extraterrestrials, to the extent of spawning a TV/movie franchise, *My Favorite Martian*.

Perhaps the greatest "disappointment" has been the lack of conflict with Martians. After over a century of speculation about canals, superintelligent

Martians, and invasions by them, not to mention numerous movies depicting them, we sent probes to Mars and found that none of it is true.

However, given various patterns in history and in STEM, we show that Martians—and a deadly conflict with them—may one day come true. Will they be little green guys and have antennae? We say no. Martians will look a lot like Earthlings, and the conflict will start with Earthlings invading Mars.

The Purpose of "Science Faction"

While this work will be sold in the nonfiction section of the bookstore, we have a high regard for the science fiction genre. Many extremely creative writers such as Isaac Asimov, Robert Heinlein, Andy Weir, James S. A. Corey, and countless others have inspired millions of readers to think about critical issues of space exploration, technology, politics, and the human experience.

This is a good thing. In fact, we encourage our readers to explore these books and think critically about their meaning, yet we purposely avoid using these works in our work. Our purpose here is to be as factual as possible and to map out what we believe is a feasible pathway for Earthlings fighting a war in outer space. To engage with the vast body of science fiction would distract us from our central thesis and lead us into the genre of literary criticism.

If we were to name a single book that influenced this work, it would be the 1996 nonfiction book *The Case for Mars: The Plan to Settle the Red Planet and Why We Must* by Robert Zubrin. This book laid out a practical and detailed plan not just for reaching Mars but for colonizing it—certainly a first step if we are to have Martians to attack. Zubrin's book fired the imagination of the older member of our team, who often used it as a resource in his classroom. Our work hopefully will extend Zubrin's ideas while adding many new elements to them.

We fully recognize that much of what we have written about will not happen or will occur in ways that are laughably different from what we suggest. This does not deter us and should not deter you, the reader. We understand and embrace the limits of writing in this matter. Like the prisoners in Plato's allegory of the cave, we are using what knowledge we have (or think we have) to better understand the universe around us.

Rather than looking at this book as a definite prediction, we want to inspire our readers to think, ask questions, and have fun while doing so. If we can do those three things, we will consider this book a success, even if all of our predictions prove incorrect and we will have inadvertently written a science fiction work of our own.

Our Three Goals: Think Critically, Ask Questions, and Have Fun!

One of the major goals of this book is to get readers thinking critically. Martian colonization is ideal for this purpose because it is still in the theoretical stages—it has not happened yet. This is useful because it frees readers of historiographic and evidentiary debates. There is no orthodox school of scholarship on Mars, there are no revisionists, and scholars do not argue about how to interpret evidence from the historical record.

This very lack of a historical background or academic literature allows us to create a world of our own. For it to have any value, our hypothetical world must be convincing enough for readers not to reject it immediately as implausible or silly. If we can get you past this hurdle, we hope to then get you thinking critically about our claims. Do they make sense? Is the logic used to support them internally consistent? In what ways could we be wrong? How would we know if we were wrong?

In an effort to encourage critical thought, we attempt to make our assumptions as transparent as possible, and we try to argue in a way that avoids jargon, complex math equations, obscure trivia, and a large number of potentially off-putting footnotes. Keeping the book as simple as possible hopefully allows the reader to focus more on the broader implications of our claims and to think critically about them. To assist the reader, the book asks simple hypothetical questions that will make readers anticipate the next part of our argument and hopefully ask questions of their own.

We believe that good books provoke more questions than they answer. Rather than simply tell you what to think, we encourage questioning. A multidisciplinary book about Mars should fit this bill nicely. It presents an argument that is novel and provides an opportunity to start the process of thinking and discovery anew.

Despite the fact that this book is about the future of Mars, many of the issues we discuss should be familiar to modern Earthlings. This is a purposeful choice. In the book we ask about the sources of wealth; the role of fear in government spending; the impacts that robots, 3-D printing, and other technologies have on people and society; the origins of revolution and war; the sources of lasting and just peace; and more. While we present these questions in the context of future Martian colonization, each of these issues is important in our current era. If we can think about why these issues matter in a hypothetical Martian colony, then perhaps we can better analyze how they might impact early-21st-century Earthlings.

While we encourage our readers to think critically about complex issues such as these, we also want them to have an enjoyable experience. From its original conception, this book was intended to appeal to educated general readers, science fiction fans, academics, military professionals, and assorted nerds. By illuminating a fascinating pathway for interplanetary colonization and war, we hope to kindle the imagination of a diverse audience.

Although this is a serious book about an important topic, we also believe that we should not take ourselves too seriously. In our experience, teaching is much more effective if it is fun. We hope that our attempts at provocation, political commentary, good-natured sarcasm, and nerd humor satisfy that need. If you can learn a little about history, politics, physics, chemistry, robots, ethics, or whatever while having a good time, so much the better!

Plan for the Book

In the broadest form, our argument is divided into two thematic sections. The first discusses how and why humans will colonize Mars and will attempt to demonstrate why a wide range of physical and political forces may lead to conflict. The second section discusses the resulting war, predicts how technology might shape the fighting, and discusses the prospects for peace and reconciliation between the two planets once the conflict is over.

Each of the chapters within these two sections addresses specific problems about colonizing Mars, effectively governing it, or the realities of warfare in space. While each of these may be read as stand-alone arguments, we have attempted to arrange them in a manner that is logical and builds on the assumptions laid out in the previous chapters.

Throughout this journey, we ask that you engage with our arguments by thinking, questioning, and having fun.

Chapter One

The Motives for Colonization: How to Not Go Broke Accumulating Money

THE SPANISH EXAMPLE IN THE NEW WORLD

History teaches that the quest for wealth (greed) is a powerful motive for placing bets on the roulette table of colonization. But have no doubt: the vast majority of players eventually lose (although for governments it may take a few centuries to do so). Indeed, casino games, regardless of type, are especially devious and addictive because they allow participants to win—just enough to give the illusion they can keep winning. Casino operators accomplish this illusion by setting the odds of losing so that they only slightly favor the house. The odds of success for colonization, especially when first established, often have been far crueler.

Human psychology has created a phenomenon called the law of small numbers, or hot hand. As applied to games of chance, it can be stated like this: If a gambler wins several times in a row, he is assumed to be a good bet for future successes. But that's not how mathematical reality works. Previous wins do not change the odds of losing on the next bet. To make matters worse, gamblers invariably will brag about their successes, thereby creating the illusion of having the hot hand even when in reality they're going broke—definitely sucker bait for other would-be winners.

Unfortunately, the law of large numbers (a foundation principle of probability and statistics) coldheartedly says that with unfavorable odds, if a participant continues to gamble, he or she eventually will lose more than was gained. This is not a matter of psychology but of mathematics. Of course, colonization is far more complex than a simple casino game. Its wagers are huge and can go on for centuries involving not just individuals but entire power structures and cultures. Nonetheless, colonization is a gamble.

To make matters worse, with colonization, the means of winning (satisfying greed) is not always clear. For example, there's been great confusion about the difference between collecting money and building wealth. The Spanish model of conquest sent groups of ruffians with a sprinkling of missionaries to the New World to collect gold and silver along with converts. In the long term, however, it did little to build wealth for Spain.[1]

The Spanish ruffians did indeed return a supply of gold and silver to Spain's royalty, but by the time the royalty covered the expenses of digging it up or stealing it from the locals, paid for the ships to transport it, built warships to protect it from pirates, and incurred various losses from things like hurricanes and theft, a significant part of the wealth was consumed before the booty reached Spain. Invariably, the booty had to be melted down and converted into coins, yet another expense, albeit a minor one compared to the others. The shrinkage was similar to what modern-day lotto winners experience when they find out they have to pay taxes.

However, the Spaniards suffered from yet another evil: inflation. When the money supply—in this case, gold and silver coins—significantly increased, a standardized unit of it became less valuable. As a result, the surrounding countries sold Spain goods and services at inflated prices. While these countries tended to prosper from the transactions, Spain did not.

In the end, neither shrinkage nor inflation bankrupted Spain. Similar to modern-day lotto winners, instead of investing in the systems and infrastructure needed to create real wealth (as opposed to money), the Spanish royalty went on a spending spree. Even worse, they borrowed more money and "invested" in what could be considered the modern equivalent of trips to Las Vegas: wars with their neighbors including France, England, and the Netherlands. By 1575 Philip II of Spain went bankrupt despite a massive influx of gold and silver.

Unlike wars of conquest in the New World, wars in Europe were fought against well-armed foes who, unlike the Aztecs, had no illusions that the Spaniards might be deities of some sort. To make matters worse, the Spaniards were not particularly innovative at analyzing what they needed in order to win, at least in the case of warships. The English were able not only to successfully harass and steal from Spanish treasure convoys, but in 1588, when the Spaniards attempted to retaliate decisively, they were dealt a horrendous disaster.

The Spanish Armada sailed with 130 ships armed with 2,500 cannons, but nonetheless depended primarily on winning by boarding their enemy's ships with overwhelming strength. Meanwhile, the English in smaller, faster, and more maneuverable ships depended on using their long-range cannon and had no intension of engaging in boarding brawls.[2] Between attacks by the English

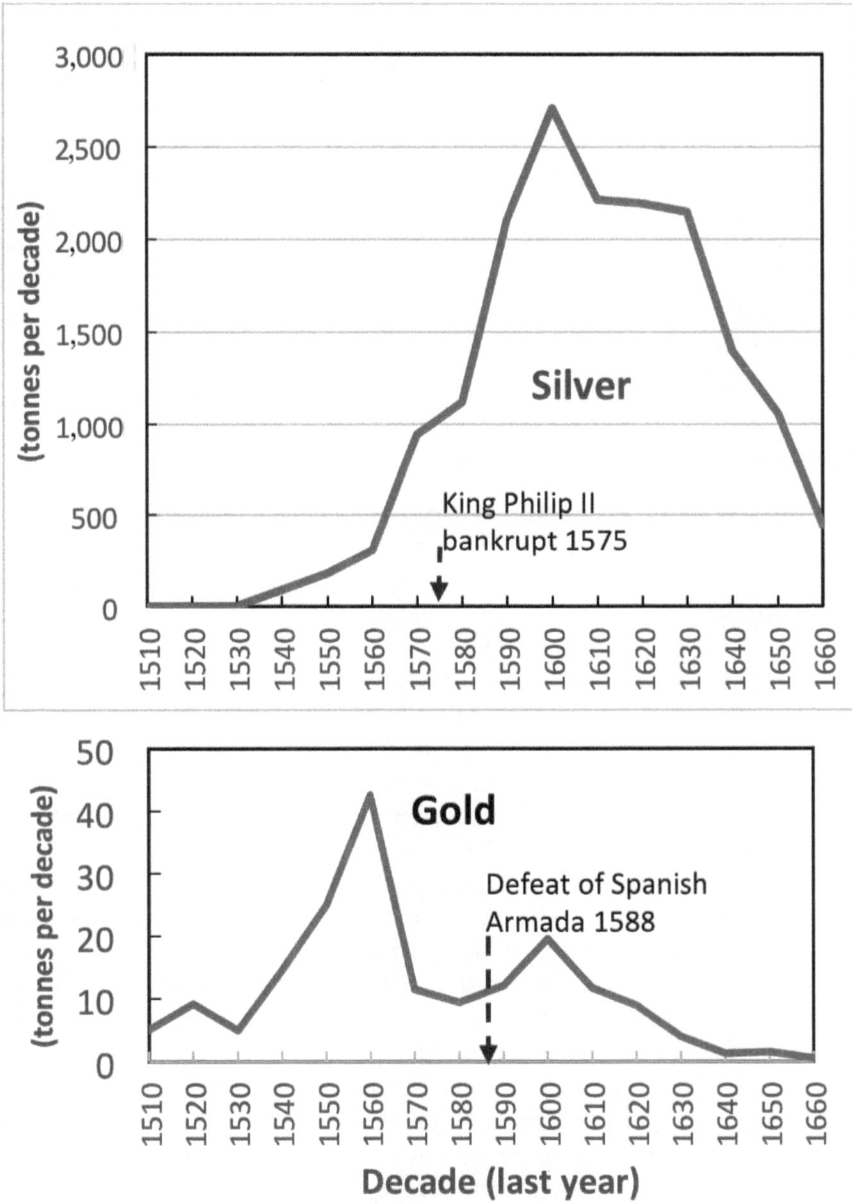

Figure 1.1. Spanish imports of precious metals, by decade, from the New World. *Source*: Author-generated graph. Data from "Gold and Silver in the New World," updated April 1999, http://mygeologypage.ucdavis.edu/cowen/~gel115/115ch8.html.

and bad luck with severe weather, the Spanish Armada lost half its ships and never fully recovered.

The precious metals from the new world did produce a real-world benefit, just not to the expected extent. When converted into coins, it became one of the world's first international monetary systems, one that significantly facilitated international trade.[3] Without actually realizing what they were doing, the Spaniards were providing a valuable service. Unfortunately, the wealth the Spanish acquired from the service of creating coins was far less than the coins' face value. So, while the Spanish did indeed benefit from the gold and silver shipped from the New World, they tended to overestimate its worth. Of course, the supply of New World precious metals also did not prove to be endless (see figure 1.1).

The Four Sources of Real Wealth

The four sources of real wealth—as opposed to money that merely represents wealth—are the availability of the items listed below. New wealth coming from these sources would be the real payoff from colonization of the next new world.

1. *Energy* is like a form of magic; nonetheless, physics tells us exactly what it can and can't do. The first law of thermodynamics tells us we can't destroy it (although we can change its form), making energy seem not just incredible, but immortal. Then, just like a scene from *The Wizard of Oz*, along comes the second law of thermodynamics and pulls away the curtain. The second law tells us that even though we can't destroy energy, as we use it, energy becomes increasingly useless. For example, if we turn on a ceiling fan, the energy used to make it spin will still continue to exist but will never again be capable of doing an activity like spinning a ceiling fan.

   In scientific terms, the energy used to run the fan ends up with an increased entropy. While still existing, after spinning the fan the energy becomes greatly dispersed inside the room, causing the temperature to get a tiny bit warmer (although the moving air makes people feel cooler). The energy is like a valuable Chinese vase that has been smashed against a concrete floor. After being smashed, the vase still exists but in a dispersed and far less valuable form. Energy loses value every time it's used. Unfortunately, using only his or her own power, a human has very little ability to rapidly change anything in the physical world. For example, a human laborer can only consistently produce about 75 watts of power as compared to a compact car, which can produce around 100,000 watts. Even a simple hair dryer can output 1,500 watts, which in a typical North

American winter could not keep a single room warm in an average-sized home. Keep in mind that power indicates how fast energy can be put to use. Many times, a human can perform the same task as a powerful machine, but it will take orders of magnitude longer to do so. Without an external source of energy humans would be weaklings shivering in the dark—at least at night in cold climates. Energy sources are profoundly enabling.

Given the relative weakness of humans and the paradox that energy cannot be destroyed but becomes worthless when used explains why finding an inexhaustible source (in human terms) arguably has become the holy grail of modern industrial societies.

2. *Useful materials* are types of matter that provide physical properties people desire. These can have useful structural properties such as lumber, concrete, or steel, useful electrical properties such as copper used in electrical wire, barrier properties such as plastic used for bread wrappers, or an endless number of other properties. Modern materials have enabled everything from smart phones to aircraft.

   The good news about useful materials is that we don't need to find them as is in nature. Given the right information about their chemistry and the right amount of energy resources, we can modify naturally occurring materials or even make synthetic materials with the desired properties—plastics, carbon fibers, paper, metal alloys, glass, and so forth. The bad news is that we have almost no ability to transform atoms from one type to another. So, for example, if we want iron atoms for making steel we have to find a supply of material with iron atoms in it and extract the iron—a process known as mining and refining.

   Of the 118 elements in the periodic table of elements, 90 can be found in nature in amounts that could be considered minable. In addition, the number of applications for these elements is generally increasing, especially for some of the more obscure and difficult-to-obtain elements such as the rare earth metals. Intel manufactured computer chips using only 15 elements in the 1990s. The number had increased to nearly 60 by 2015.[4]

   We are often walking around atop the very atoms we strongly desire to use. The problem is that they are greatly diluted with other atoms and thus hard to recover. Finding concentrated sources of useful elements would be worth some colonization effort.
3. *Information* is any form of symbolism, data, knowledge, instructions, recipe, or algorithm that serves a purpose desired by humanity. It includes both art and science. Information controls and shapes all the other forms of wealth. Without it, they are useless.

While information is stored in many types of physical objects such as paintings, books, and hard drives, the storage form is merely a container that by itself does not have the same value as the information it contains. For example, oil paints have relatively little value, but arranged on a canvas by Rembrandt, the paints contain a priceless form of information.

As we have entered the digital age, we have begun to mathematically and scientifically characterize information. A quantity of information is now often express by a number of bits. Thanks in large part to the ground-breaking work of Claude E. Shannon, considered the father of information theory, we also have mathematically demonstrated that information like energy, has entropy (essentially a measure of uncertainty).[5]

Although much of our mathematical analysis of information is new, the high value placed on it has been reflected in numerous systems designed to protect it, including such things as patents, copyrights, trademarks, encryption, and closely held trade secrets such as the formula for the soft drink Coke. Indeed, stealing or revealing protected or secrete information can lead to severe civil and sometimes criminal penalties. In the case of military secrets, it potentially can get the offender executed for treason.

Unlike the other forms of wealth, information can be created by thought, so why go to all the bother of colonizing a new world simply to gain information? Why not just sit under a tree and think about it? It's not that simple: Mathematical modeling and computer simulations all can help, but in the end, information depends on sensory input. In fact, anything that expands our sensory ability (telescopes, microscopes, slow-motion videos, instrumentation) increases our available information. Exploring a new world would pretty much top the list, both from the standpoint of new scientific discoveries on Mars and from the standpoint of facilitating future space exploration.

4. *Territory,* especially its subset of habitable space, is essential for human survival. It's typically measured in units of area with the implicit assumption that there is at least enough headroom above it for reasonably tall people to do standing jumps without hitting their heads. Hence, a skyscraper will contain many times more area than its footprint on the ground. The value of territory can be a challenge to quantify with scientific and mathematical principles. Still, all things equal, spacious houses and apartments are more valuable than smaller ones. Wealthy people often own multiple dwellings that are anything but tiny cottages, while poor people sleep under bridges. First-class seats in airliners are wider than the less-expensive ones. Clearly, the territory under one's control is an indicator of wealth.

It is also clear that a certain amount of territory is required to take care of people's physical needs other than shelter. These needs include water, food, clothing, and waste disposal. All other things being equal, a greater land area on Earth will collect more solar energy and rainwater, be able to support more plant and animal life for providing human needs such as food, and provide more options for dealing with waste disposal.

Since territory is symbolic of wealth, it has considerable emotional as well as physical value. For royalty and despots, conquering territory has often been a top priority. Acquiring additional territory or living space was one of the reasons Adolf Hitler invaded eastward in World War II. Of course, taking the Baku oil fields—a much needed energy source for keeping his battle tanks running—was also a factor.

How the Sources of Wealth Interact

The four sources of wealth are not isolated categories. For example, food is a material, but as a source of wealth, it rightfully falls in the energy category, since it is the energy source that powers humans. Water, on the other hand, is a useful material because although essential for human life it does not fuel humans. When used as a fuel, oil falls in the energy category. When used as a lubricant, it falls in the useful-materials category.

What seem like relatively modest differences in the availability of materials and the know-how (information) required to use them for creating useful products can have a remarkable effect on everything from standard of living to military might. On November 16, 1532, Spanish conquistador Francisco Pizarro, along with 106 soldiers and 62 cavalrymen, captured the Inca emperor Atahualpa after sending the Inca army of about 80,000 soldiers fleeing in terror. By the end of the day, the Spaniards had killed about 7,000 Incas—a number limited only because darkness fell—without a single loss of their own. Of course, the fear induced by the use of horses and some crude early firearms (both of which can be viewed as energy sources) was a definite factor in the victory—the main factor, however, was the know-how for manufacturing (information) and the subsequent availability of a single material, steel, used for fabricating the Spaniard's helmets, armor, and weapons including their firearms.[6] In other words, conquest was largely made possible by a material that used a combination of two elements from the periodic table—iron and a whiff of carbon. Likewise, in modern times a significant military advantage could be incurred by something as obscure as a new application of a rare earth metal in an electronics system for something like a guided missile or self-guided projectile.

In the end, Atahualpa gave the Spaniards a treasure trove of gold as a ransom for his life, an agreement not honored by the Spaniards who eventually executed him. Given the outcome, one cannot help wondering which was truly more valuable, the Inca's gold or the Spaniard's steel.

In preindustrial societies, the energy required for building essentially all public works, including the pyramids, Greek temples, Roman roads, and Gothic cathedrals, was provided by a combination of people and animals, which in turn were powered by crops that required a vast amount of territory. It has been estimated that just the energy needed to build the Roman Colosseum required about five years of output from an area of crop land about the size of Manhattan Island.[7] Add the fact that firewood often has been humanity's primary source of heat, and it's clear that territory (actually the solar energy shining on it collected by various plants) has, in a sense, been the primary source of energy for most of human history. This helps further explain why acquiring additional territory has been such a big deal for conquerors and tyrants. To gain territory was, essentially, to gain power in the physical sense.

The territory required for powering humans also has been greatly influenced by information. Depending on the lushness of the landscape, hunter-gatherer cultures using little to no agricultural information required roughly 600 to over 34,000 acres of land to feed or, in other words, power a human.[8] By contrast (while estimates vary), in a temperate climate only about an acre of land is needed to power a person using the best available farming information.[9]

Information has had a profound influence on the efficient use of all resources. Historically, though, the quest for new information has not been a significant motivating factor in colonization. But in recent times, public spending on space exploration, officially speaking, has been based on the quest for new information—in other words, scientific discovery. The return on investment (ROI) has taken two forms: new scientific knowledge and the spinoff products resulting from the development of new technology to meet the challenges of exploration. Although there are essentially no modern devices containing microprocessors or modern materials that have not in some way benefited from space technology—benefits to everything from aircraft to cell phones to computers—the public in the United States has been double-minded about funding space exploration.

The Interaction of ROI and Transportation Costs

Of course, wealth building through science has not been the only motivation for politicians to spend money on America's space programs. Political maneuvering for things like a military advantage, a means of fostering international cooperation, or simply a means of distracting the public from embarrassing political failures also has been a factor. Yet who cannot help being inspired by John F. Kennedy's famous speech in which he proposed going to the Moon not because it is easy but because it is hard? What other multibillion-dollar program could be justified "because it is hard"?

In its day, the attitudes toward the space race (culminating in the Apollo missions) were either giddy support or total disdain at "wasting" money on rockets while there were people living in poverty. In the beginning the first attitude dominated. It was our team, USA, USA, USA versus the Russians—football in the sky. And at halftime we were losing. The Russians were the first to put a satellite in orbit, the first to blast a man into space, and the first to orbit a man around the Earth. Then came the magical moment: touchdown—a man for the first time ever walked on the Moon. Our team had won, and the crowd went wild! Subsequently, the Apollo program was canceled with hardly a thought about returning, let alone establishing a lunar colony.

The sustained effort needed for lunar exploration, let alone colonization, had been killed by transportation costs. Once the giddiness of winning the space race wore off, launching three guys into space atop a rocket weighing 62 million pounds, 36 stories high, six different times in return for 382 pounds of moon rock, with no hope of reusing it, did not look like a way to feed the poor, house the homeless, or for that matter enrich the wealthy.

For the would-be space colonist, the minimum bet at the roulette table of colonization is the cost of transportation to the colony (assuming the colony is well enough developed so that immigrants can trade labor in exchange for room and board on arrival). Buying a third of what could be considered a super high-tech 36-story disposable skyscraper just to get there, not to mention the cost of fuel, is not going to be enticing.

For the investors, the minimum bet will include the cost of multiple back-and-forth trips. For either the investor or would-be colonist, an extortionately high transportation cost ends the venture before it begins. But neither investors nor immigrants will have to pay full price for the minimum bet. It will be subsidized by public funds. NASA is already paying private space exploration companies to launch scientific-related payloads. This allows the companies to develop and perfect their space-travel capabilities and infrastructure while earning a profit. In the end, NASA will directly fund at least some of the colonization effort as well as buy services and products from the space exploration companies. Part of colonization will be seen as a jobs program, not to mention that the U.S. government is unlikely to walk away from the creation of a new world and abdicate all possible influence over it.

Nevertheless, the most important boost to space colonization will not come from government subsidies but from substantial decreases in the cost of transportation. The cost of going into space from the early days of NASA to the current efforts of private-sector companies like SpaceX has fallen by orders of magnitude and continues to fall. Private-sector companies are doing this in part by eliminating much of NASA's bureaucratic approach to the design and launch of space vehicles. Computer-aided design (CAD), computer-aided manufacturing (CAM), and over 70 years of rocket science development also have helped.

However, the use of reusable vehicles is probably the most promising cost reducer. SpaceX already has successfully launched rockets and then recovered the first-stage boosters by landing them in an upright position using rocket power to slow the descent. This new technology is currently focused on launches to low Earth orbit and already saves roughly the cost of a jumbo jetliner on each launch. Eventually, manufacturing economies of scale for launchers will further reduce costs.

The NASA Space Shuttle attempted to use this same reuse strategy for reducing launch costs, but at about $10,000 per pound sent into low Earth orbit, the Space Shuttle largely failed to deliver.[10] It was like trying to design an off-road vehicle with the carrying capacity of an 18-wheeler and the performance of a Porsche. The shuttle was not only massive but massively complex, which greatly extended the time it took to prepare it for each flight and mired it in bureaucracy. Losing a shuttle would result in a very visible loss of human life, not to mention a loss of billions of dollars—a bureaucrat's worst nightmare.

The shuttle's reusability actually worked against it. It sucked up so much budget just to operate that, given its size and complexity, it could not evolve. There seems to be an optimum size for cost-effective space-launch vehicles, and the shuttle was larger than the optimum point for the era's technology.

As travel to space colonies matures, emigrants from Earth either will be launched in large spacecraft atop huge rockets into low Earth orbit where they will be refueled for the rest of the trip to Mars, or they will be launched a few people at a time and collected in a much larger spacecraft, which will then complete the trip to Mars. These larger spacecrafts will have been built in and permanently remain in outer space. They will be made from materials acquired in space, thus avoiding the horrendous cost of launching from Earth. With either option, all elements of the spacecrafts will be used multiple times, thus making the cost of travel far lower than in the early days of exploration.

Given highly developed space travel technology and some level of government subsidies, the price of a one-way ticket to a new world such as Mars could be reduced to about the cost of an Ivy League education. Unfortunately, the tired, poor, huddled masses yearning to breathe free need not apply. Not only is the price of a ticket going to be high, but earning a living on Mars is going to require a high level of training, education, and/or skill. The cost of transportation will be a significant barrier to the colonization of new worlds, but not the only barrier.

Why a Martian Instead of a Lunar Colony

Unlike the previous space race, the next race will focus on colonization, not of the Moon but of Mars. Although the Moon is closer, it takes almost as much fuel to reach it as to travel to Mars. The Moon has only 16.5 percent as much gravity as Earth, giving no hope of ever having a significant atmosphere. On the other hand, Mars's gravity is about 37.8 percent as high as Earth's, and with terraforming, Mars could have an atmosphere with a pressure of around 5 psi. Keep in mind that Apollo astronauts breathed pure oxygen at 5 psi when in space.[11] So even if the outside atmosphere were not breathable due to low oxygen content, the pressure inside a Mars habitat could be the same as outside and still be breathable with pure oxygen. Matching the inside and outside pressures would greatly reduce the possibility of structural failure and simplify construction.

With such an atmosphere, Martian colonists could walk around outside without wearing bulky pressure suits (although they would still need breathing gear). The atmosphere also would offer some protection from cosmic radiation and solar flares—a benefit not available on the Moon.

Lunar nights last about two weeks, making solar cell panels essentially useless while also complicating possible agriculture. Given the lack of an atmosphere and the length of the night, the Moon is incredibly cold at night and hot in the day. By contrast, the length of the Martian night is roughly the same as Earth's, and with terraforming, Mars would be positively balmy compared to the Moon.

The Irrational Motives for Colonization

The colonization of Mars offers the potential for a significant payoff in all four areas of wealth (as will be discussed in subsequent chapters), and the cost of getting there will be within reach, but its implementation will involve, if not require, some romanticized delusions. For the billionaire funding private space exploration and colonization, owning a private island will be so yesterday compared to the idea of owning a private planet. For the engineer or computer geek, creating a new world—a real one, not just a cyber one—compared to spending one's life as a nameless soul trapped in a cubicle will be positively seductive. For the scientist who faces a lifetime of seeking a new discovery that may never be realized, the prospects of walking on a planet where one can make a new discovery every day will be irresistible.

Well-established settlements or worlds can produce a level of comfort that seemingly should remove the desire for colonization, but in doing so can unknowingly create conditions that make emigration attractive. The economist Vilfredo Pareto observed in the early 20th century that a small percent of a country's population invariably will control a disproportionately large percentage of its wealth. Wealth incurs political power, and political power is often used for creating rules and laws that preserve wealth with the side effect of restricting upward mobility. Even rules with the best of intentions tend to favor wealthier established businesses because they are better able to cope with rules than much smaller new ventures. A large and crowded population intensifies disparities. These things make the prospects of a new world alluring. Furthermore, biology itself imparts a drive for survival of the species that for humans seems to translate into a desire for expansion into new territory. Get Mars viable for settlement, and it would attract nerds, entrepreneurs, and adventurers like a magnet.

Romanticized delusions have always been a key motivation for people to give up security and relocate to new worlds. Furthermore, such delusions were not limited to Spaniards. The first English settlement in America, Jamestown, was established in 1607 by 104 settlers, 48 of whom were gentlemen.[12] These individuals were upper-middle-class men, just a step below royalty, who had paid to join the venture. One wonders, what were they thinking? Certainly, not that after arrival, they would soon be dying of disease and that those lucky enough to survive and remain until "the starving time" in 1609 would not just be eating their cats, dogs, and horses but apparently also be eating their fellow colonists.[13]

Selling would-be colonists the idea of emigration to Mars and convincing investors and politicians to fund it will definitely involve romanticized delusions, but when it comes to motivation, if the price is right, nothing, not even greed, beats romanticized delusions . . . that is, except for fear.

NOTES

1. See generally J. C. Sharman, *Empires of the Weak: The Real Story of European Expansion and the Creation of the New World Order* (Princeton, NJ: Princeton University Press, 2019), 39–47.

2. Daryl Worthington, "Spanish Armada Sets Sail From Corunna," accessed June 5, 2019, https://www.newhistorian.com/spanish-armada-sets-sail-from-corunna/4360/.

3. Peter Gordon and Juan José Morales, "When the Dollar Spoke Spanish," accessed June 5, 2019, https://www.asiaglobalonline.hku.hk/when-the-dollar-spoke-spanish/.

4. David S. Abraham, *The Elements of Power* (New Haven, CT: Yale University Press, 2015).

5. Graham P. Collins, "Claude E. Shannon: Founder of Information Theory," accessed June 5, 2019, https://www.scientificamerican.com/article/claude-e-shannon-founder/.

6. Jared Diamond, *Guns, Germs, and Steel: The Fates of Human Societies* (New York: W. W. Norton & Company, 1997).

7. Thomas Homer-Dixon, *The Upside of Down: Catastrophe, Creativity, and the Renewal of Civilization* (Washington, DC: Island Press, 2006), p. 53.

8. Ronald L. Conte Jr., "Hunter-Gatherers No More," Hunger Math, accessed June 5, 2019, https://hungermath.wordpress.com/2013/11/05/hunter-gatherers-no-more/.

9. Jason Bradford, "One Acre Feeds a Person," accessed June 5, 2019, http://www.farmlandlp.com/2012/01/one-acre-feeds-a-person/.

10. David Kestenbaum, "Spaceflight Is Getting Cheaper. But It's Still Not Cheap Enough," accessed June 5, 2019, https://www.npr.org/sections/money/2011/07/21/138166072/spaceflight-is-getting-cheaper-but-its-still-not-cheap-enough.

11. Amy Shira Teitel, "Why Did NASA Still Use Pure Oxygen After the Apollo 1 Fire?," *Popular Science*, accessed June 5, 2019, https://www.popsci.com/why-did-nasa-still-use-pure-oxygen-after-apollo-1-fire.

12. National Park Service—Historic Jamestowne, "The First Residents of Jamestown," accessed June 5, 2019, https://www.nps.gov/jame/learn/historyculture/the-first-residents-of-jamestown.htm.

13. Joseph Stromberg, "Starving Settlers in Jamestown Colony Resorted to Cannibalism," accessed June 5, 2019, http://www.smithsonianmag.com/history/starving-settlers-in-jamestown-colony-resorted-to-cannibalism-46000815/.

Chapter Two

The Boogeyman in the Dark: Fear of Flying Rocks

When it comes to loosening the purse strings on governmental spending, nothing beats a boogeyman hiding in the dark—especially one hiding in the darkness of outer space. While a Martian colony will attract adventurers, scientists, and speculative investors, without some level of governmental spending to support the endeavor, the massive sunk costs of capital and the decades-long time frame to recoup the initial investment might sabotage the exploration and colonization of Mars. Fortunately, some level of government spending will be available from organizations such as NASA. Indeed, as major companies become increasingly committed to the colonization effort, they will likely exert political influence to maintain or increase governmental support.

Based on recent history, politicians would invest, at least to some extent, in colonization as a prestige project (much as the United States and Soviet Union did during the Moon race), but doing so could easily morph into a massive and unpopular financial commitment. Without a clear benefit for taxpayers, the endeavor probably would be one of the first items cut when facing budgetary uncertainty or citizen demands for tangible benefits from their tax dollars. In fact, this is just what happened to NASA after it won the race to the Moon. Although the United States triumphed over the Soviets, and generated a plethora of spinoff technologies and products in the process, not much of immediate value was found on the Moon itself. Subsequently, the government reallocated its limited resources, ironically, on various wars targeting various boogeymen: the war on poverty, the war on inflation, the war on drugs, the war on cancer, and the war in Vietnam.

Unlike the space race or fears of a missile gap during the Cold War or the scramble for Africa during the 19th century, the longtime horizons for colonizing Mars, in addition to the financial costs, could make massive government spending unsustainable. In each of these previous competitions, there was a clear goal (beat the Soviets to the Moon, achieve superiority in ICBMs, take the richest parts of Africa), a considerable time pressure for reaching the goal, an important benefit (prestige, avoiding nuclear destruction, and access to the rubber and ivory trades), and definite winners and losers. Without some driving sense of urgency or fear, it's highly unlikely that a government will be able to sustain a high level of spending on the Mars effort. And there just happens to be a lurking boogeyman that could ignite a virtual firestorm of federal spending.

FEAR AS A DRIVER FOR MARTIAN COLONIZATION

Should such a boogeyman (a dangerously large space rock on a collision course with Earth) suddenly appear from the darkness of outer space and do grievous harm—the kind that provides numerous videos of mangled bodies, bloodied faces, and burned flesh—politicians would jump at the chance to fund preventative measures. And, in the case of a meteorite strike on a populated area, these preventative measures would of necessity involve the types of technology used in space exploration, asteroid mining, and space travel—the very technologies required for developing and maintaining a Martian colony. Potentially dangerous asteroids in outer space are that perfect boogeyman for motivating government spending: They are really scary and are a problem that will never completely go away.

While environmental factors such as global warming and demographic pressures such as overpopulation are perhaps even more serious than meteorite strikes, these issues will take decades, possibly centuries, to threaten human existence. This slow slip into a bleak Malthusian future is so gradual, in fact, that few people lose much sleep about the future existence of life on Earth.[1] Moreover, because these long-term problems will only incrementally degrade the livability of Earth, there is still time to implement wedge strategies to lessen the impact of greenhouse gases or bend the population curve to a more sustainable growth rate. This is not to suggest that global warming and population growth are not threats, but because they develop over many years, they do not cause the type of fear and panic needed to induce an immediate dramatic revision of government spending and policy.

This is analogous to a wrinkle on your face. You do not suddenly wake up one day, look in the mirror, notice a wrinkle, decide that you are old, and choose to give up smoking and become a vegan. The wrinkle develops almost imperceptibly over time, and you never have a single moment where you realize that your youth, decent looks, and athletic abilities are gone.

Heart attacks, on the other hand, are a shock to the system. Heart attacks make people reevaluate their lifestyles, stop smoking, go vegan, find religion, and make amends with estranged family members. Assuming that you wake up at all, this shock to the system is a wake-up call that actually causes people to change in a way that a wrinkle or losing your touch on your jump shot never would.

To gain significant sustained governmental support for the colonization of Mars, mankind needs a "heart attack" moment—big enough, shocking enough, and sudden enough to demand our attention, but small enough to survive and retain the ability to change. The most likely heart attack for Martian colonization is a meteorite strike—like the one that killed the dinosaurs, but not *that* bad.

The Problem of Meteorite Strikes

Every day, literally thousands of objects from space enter the Earth's atmosphere. Most are tiny, about the size of a speck of dust. These burn up in a fraction of a second and are noticeable only with high-tech detection equipment. Some are larger, up to a meter across, and are visible to the naked eye as they enter the atmosphere. Traveling at speeds of up to 30,000 miles per hour and reaching temperatures of about 30,000 degrees Fahrenheit, these "shooting stars" are beautiful but typically harmless as they combust and are completely consumed by the heat generated by their passage through the sky.

A very small percentage of the thousands of objects that enter the atmosphere ever reach the ground. These extraterrestrial bodies, known as meteorites, strike with an enormous release of explosive energy after impact with the surface, typically leaving craters 12–20 times the size of the meteorite itself. The exact size and destruction of a meteorite depends on many factors such as its physical composition, the angle at which it strikes the Earth, its temperature, and the type of terrain it impacts. The kinetic energy contained in large-sized meteors is almost instantaneously released as a massive explosion when it hits the ground. The explosion is so intense that it can actually vaporize the meteor. Needless to say, even a small meteorite can impart tremendous damage to whatever it strikes.

However, large-sized meteors can also get so overheated that they explode in the atmosphere before they hit the ground. This even depends on the composition of the meteor and is more likely to happen to stony-type rather than metal-type meteors. Nevertheless, even these explosions can be devastating.

While very few people lose sleep at night worrying about meteorite strikes or the fate of the dinosaurs, this may be the result of ignorance, not science.

In fact, major meteorites do strike the Earth's surface or explode just over the surface with a frequency of about once every 60 years. Given their massive destructive potential, it is worth considering two of the most prominent recent examples.

On June 30, 1908, a massive meteorite exploded in the atmosphere as it headed toward the ground near the Tunguska River in Siberia. Thousands witnessed the event as it traced through the atmosphere, but fortunately none were believed to have been killed by the impact, thanks to the fact that it came down in a remote and unpopulated wilderness area. The meteor is believed to have had the explosive energy of approximately 15 megatons of TNT or approximately 1,000 times the destructive energy of the atomic bomb dropped on Hiroshima!

The meteorite leveled over half a million acres of forest and illuminated the night sky in the area for several days afterwards. Had the meteor been a different composition—a metal instead of a stony type—it actually could have reached the ground before exploding, creating a massive crater that caused even more damage.[2]

The "Tunguska Event" was not scientifically investigated until 1921, and debate continues to surround it to the present day. Indeed, the mysterious nature of the event, combined with a lack of scientific consensus and a mistrust of Soviet accounts, has fueled wild speculation about the incident. It is the plot point for dozens of novels, comic books, and films; some have asserted that it is evidence of an alien attack on Earth or the crash site of a space ship from another world. While these claims are clearly bogus, the fact that this incident is still remembered today speaks not only to the destructive power of the meteor, but also to the power such an incident has on popular imagination.

A similar event occurred on February 15, 2013, when a 17-meter–wide rock broke apart and exploded some 12–15 miles over the town of Chelyabinsk, Russia. The explosion released destructive energy in excess of 470 kilotons of TNT, or some 30–40 times the blast of the Hiroshima bomb. Fortunately, there were no reported deaths, thanks again to the extreme altitude of the explosion and the somewhat remote location of the event. Despite this, the destruction was significant. More than 1,200 people sought medical treatment for their injuries, more than 7,200 buildings were damaged, and the economic cost for cleanup and rebuilding was estimated at over $33 million USD.[3]

Due to the many dramatic images and the extensive damage, this incident quickly captured the attention of the world press, politicians, and scientists. Russian Prime Minister Dimitry Medvedev confirmed that a meteor had struck Russia and claimed that the "entire planet" was vulnerable. In his official statement, he stated that the entire world was vulnerable and needed to create a defense system to protect mankind.

As a result of this increased attention, the United Nations Office for Outer Space Affairs accelerated its work on studying the problem, work begun earlier the same year after the near-miss of the 2012 DA14 asteroid.[4] In the weeks that followed, some scientists even proposed active defense systems using directed energy to destroy incoming meteorites, proposals that would have seemed ridiculous even a few days before.

Despite this significant media attention, very little of substance was actually done to solve the problem or alter policy in any meaningful way. Public attention quickly shifted to other issues such as the government shutdown, continuing revelations from Edward Snowden, changes in gay marriage laws, and the troubled implementation of Obamacare. Given these seemingly more pressing issues, the fear of meteorite strikes was quickly forgotten.

While NASA would spend a small portion of its limited budgets to study the problem and the UN eventually would declare June 30 as International Asteroid Day, very little of substance has been done to address meteorite-strike prevention.[5] Despite this seeming indifference, the issue has not gone away, and the Earth remains potentially vulnerable. Indeed, had the 2013 meteorite struck in a heavily populated area and killed thousands or millions of people, the narrative would have been completely different. Instead, it was quickly overcome by other events in the news cycle.

Although the big meteor strikes are terrifying at a probability of only one strike every 60 years, they could easily continue to miss heavily populated areas for a long time. However, about two meteors with a range of energy from about 1 to 20 kilotons of TNT impact the Earth every year—a probability about two orders of magnitude more likely to happen. The casualty count would be lower than from a larger meteor strike, but nevertheless an explosion of 1 to 20 kilotons of TNT in a densely populated area would create thousands of deaths and injuries.

With respect to an asteroid strike in a populated area, time is not on Earth's side. Considering that the colonization of Mars will take many decades during which the populated area on Earth is ever increasing (yielding a larger target), given the time span involved it is likely that an asteroid will strike a populated area well before Mars is fully colonized and becomes independent.

Remembering the outcry for spending triggered by 9/11, even a smaller meteor strike in a populated area would have a dramatic effect. After seeing the dramatic images, for the first time masses of people will have the thought that this "could happen to me." Human imagination will, hence, create a new type of boogeyman.

Fear, Panic, and Defense Spending in the Digital Media Age

If the threat from extraterrestrial bodies is real, then why is it not seen as a pressing global issue? The simple explanation is that the vast majority of people have not experienced it either firsthand or vicariously, at least not yet.

So far, every modern impact site of meteors has occurred in an isolated area with no significant loss of human life. If a population center were the point of impact, however, the results would be catastrophic. For example, if a 15-megaton explosion occurred in central Manhattan, the resulting blast could kill in excess of 4 million people and injure another 3.4 million.[6]

In these days of social media and camera phones, such a catastrophic event would send shockwaves around the globe. Almost instantly, gory photos of burning buildings, panicked people, and mass casualties would be transmitted across the internet, and despite government efforts to restore calm and control the media narrative, the story quickly would take on a life of its own. In fact, nearly everybody near the event would become an instant journalist, sharing graphic images and personal stories and spreading terror worldwide in a matter of minutes.

The effect of this access to instant information about the unfolding catastrophe would be contagious fear on an international scale, a mix of rumors and misinformation that could quickly incite global hysteria. Unlike the 1908 and 2013 impacts in Russia, the results would be impossible to ignore as the whole world would be watching in horror and confusion.

The shock, fear, and devastation of an asteroid striking a major city would almost certainly be a wake-up call for the citizens of Earth. Assuming that humans did not suffer the same mass extinction of the dinosaurs, they would demand action from their leaders, who in turn would be happy to throw money at anything that made them look like they were doing something.

Yet again, history serves as a guide. When discussing the 1947 defense budget, Senator Arthur Vandenberg (perhaps apocryphally) suggested that President Harry Truman should go before Congress and "scare the hell out of the American people" in order to ensure a commitment to higher defense budgets.[7] This tactic worked. Americans were scared at the inflated Soviet threat and demanded action from their leaders. The result was that, despite the financial sacrifices, large defense budgets designed to contain the Soviet threat became a bipartisan issue for the remainder of the Cold War. Because the prospect of Communist world domination scared the American people, the money flowed into the national security apparatus, and few seriously questioned the need for such increased spending.

Similarly, in the post-9/11 era, the American people have remained scared of terrorism. Defense spending and homeland security measures have become virtually unquestioned by either party, despite the fact that some if not many have a dubious value. Indeed, the United States has spent trillions of taxpayer dollars on measures such as transportation security and wars in Iraq and Afghanistan based in large part on fear generated by a few dramatic attacks.[8] While the total body count from those attacks has been horrifying, it is actually low compared to other forms of violent death, such as auto accidents.

Simply stated, scared people spend money to make themselves feel safe, even if this is irrational in strictly economic terms. While the September 11 attacks were a major global event, they would pale in comparison to the destruction of a major city by an asteroid. People would demand safety and security and would suddenly be willing to invest enormous sums in anything that could possibly provide it.

Governments also may look to private corporations to help them in their planetary defense efforts. With the government pumping dollars into them, Mars transportation companies suddenly will find themselves in a far better financial condition and be better positioned to pursue their Mars activities. Even now, SpaceX's cash flow essentially is coming from selling transportation to the International Space Station because NASA has no rocket transport capability. Without this infusion of revenue, SpaceX would be severely limited in its activities even if it were able to continue existing. Being good capitalists, such private corporations would at least attempt to get a share of the massive amounts of money stemming from a major meteorite event and may seem like a logical and expedient solution to the problem.

The result could be the governmental funding boost that accelerates Martian colonization on the path to becoming a success. Once Martian colonization becomes a security issue, it is no longer just a science project or a speculative venture by a few billionaires, but a true national and international priority.[9] With the financial backing of governments desperate to reassure their people, budgetary limitations and profit motives become largely irrelevant. When people demand action, the money will flow because the meteorite "scared the hell out of the American people."

Earth Versus Mars: Why a Mars-Based Early Warning and Defense System Makes More Sense

The first step in deterring a dangerous asteroid is to detect it—you cannot kill what you cannot see. Given Mars's position near the asteroid belt, its lower gravity (meaning easy-to-make bigger telescopes), and its relatively nonexistent atmosphere (meaning its telescopes work better), Mars will in many instances be better able to track the positions and orbits of asteroids and meteors, especially the small ones, than can be done on Earth.

Depending on where Mars is located relative to Earth when a dangerous asteroid is detected, Martians also may be better able to respond to the problem than Earthlings. Martian space mining projects already will be steering water-bearing asteroids and comets so that they deliberately hit Mars in order to heat up the planet and increase its water content. They will be the only ones with this technology, not just on paper but in actual practice. If you can cause an asteroid to hit a planet, you can cause it to miss a planet. Martians likely also will have demonstrated the technology to move small asteroids closer to Mars for mining purposes.

Unlike movies in which astronauts travel to aberrant asteroids and blow them up with nuclear bombs, the asteroid defense will be entirely automated and done with robotics that might attach ion-thrust engines to the asteroid. Some level of remote monitoring and control over this operation would be desired. Obviously, if Mars is closer to the asteroid, the time to send and receive signals to it would be reduced.

While the initial efforts at planetary defense likely would focus on Earth-bound solutions, the technical limitations discussed above would soon shift the focus to an extraterrestrial solution. Except for the greater cost and setup time, a base on Mars would be more effective in every other way than a base on Earth.

Unlike Earth, Mars does not have a significant atmosphere to obstruct the view into space. This would allow a much deeper and clearer look into space than a system based on Earth. While it would be possible to achieve similar results with space-based telescopes, these would be limited in their field of vision because they would be orbiting around the Earth and would have temporary gaps in their coverage. If there were no human activities on Mars at the time of the meteor strike, politicians would not look to it as part of the defense system for Earth. However, with the colonization and terraforming of Mars in progress, the planet would be a logical part of Earth's asteroid defense system because of its technological and positional advantages.

A meteorite strike in a populated area is a heart attack moment that uncovers the existence of a lurking boogeyman. Once the first boogeyman is actually observed, the fact that there are undoubtedly other boogeymen lurking in the darkness of outer space is enough to sustain governmental spending.

NOTES

1. See Thomas Malthus, *An Essay on the Principle of Population* (London: J. Johnson, 1798).
2. Luca Gasperini, Enrico Bonatti, and Giuseppe Longo, "The Tunguska Mystery 100 Years Later," *Scientific American*, June 2008.

3. Russia Times, "Meteorite Hits Russian Urals: Fireball Explosion Wreaks Havoc, Up to 1,200 Injured," accessed June 5, 2019, https://www.rt.com/news/meteorite-crash-urals-chelyabinsk-283/; and Moran Zhang, "Russia Meteor 2013: Damage to Top $33 Million; Rescue, Cleanup Team Heads to Meteorite-Hit Urals," accessed June 5, 2019, https://www.ibtimes.com/russia-meteor-2013-damage-top-33-million-rescue-cleanup-team-heads-meteorite-hit-urals-1090104.

4. United Nations Office for Outer Space Affairs, "Near Earth Objects and Planetary Defense," accessed June 5, 2019, http://www.unoosa.org/documents/pdf/smpag/st_space_073E.pdf.

5. Ibid.

6. Nuclear Security.com, "Nukemap V. 2.61," accessed June 5, 2019, http://www.nuclearsecrecy.com/nukemap/.

7. See generally Lawrence Haas, *Harry and Arthur: Truman, Vandenberg, and the Partnership That Created the Free World* (Lincoln, NE: Potomac Books, 2016), esp. 291n2.

8. See generally: John Meuller, *Overblown: How Politicians and the Terrorism Industry Inflate National Security Threats, and Why We Believe Them* (New York: Free Press, 2006).

According to one recent estimate, the wars in Iraq and Afghanistan have cost nearly $6 trillion USD since 9/11. See Watson Institute for International and Public Affairs, "Costs of War," accessed June 5, 2019, https://watson.brown.edu/costsofwar/.

9. On the topic of securitization, see: Barry Buzan, Ole Weaver, and Jaap de Wilde, *Security: A New Framework for Analysis* (Boulder, CO: Lynne Rienner Publishers, 1997).

Chapter Three

The Gravity Well: Defining the High Ground

The reason that Mars is vastly superior to Earth as a platform for space travel can be summarized in a single sentence: When it comes to sending vehicles into outer space or monitoring activity in the solar system, Mars essentially occupies the high ground when compared with Earth.

Even 2,000 years ago Sun Tzu recognized the military advantage offered by occupying the high ground and the disadvantage of attacking an enemy who got there first.[1] While Sun Tzu didn't elaborate on all the advantages, basic physics indicates that an arrow shot downhill would go further than one shot uphill. Soldiers charging downhill would be faster and less fatigued when they engaged the enemy than those charging uphill. In physics terms, individuals at a higher elevation have a higher gravitational potential energy, energy that could be converted into kinetic energy or the energy of motion. Essentially, this higher potential energy would make an army more powerful and more mobile, while the improved visibility from high ground made the commanders better able to collect battlefield information.

While there is no up or down in outer space, the high ground primarily can be defined by one's relative position in what could be called a gravity well.

If a long enough ladder could be raised into the heavens, as people climbed it (assuming they were properly attired in space suits) the gravity force from their planet would get less and less. At some point, the force would become so negligible that they could jump off the ladder without falling. It would be as though they had climbed from confinement at the bottom to freedom of movement at the top of a very deep well, hence the term gravity well.

For a military force to attack another planet it would first have to "climb" out of its own planet's gravity well. On planets with a stronger gravity force, the gravity well would be deeper than on a planet with a weaker gravity force.

The military force on a planet with the shallowest gravity well essentially would be on the high ground. It would use far less of its energy resources in attacking an enemy's planet than its enemy would need to attack.

This is not to say that climbing out of their planet's gravity well is the only energy requirement for Earthlings to reach and attack Mars (or for the opposite of Martians attacking Earth) but it is going to be the most significant requirement. The total energy needed will depend on a host of variables such as how fast the attackers want to arrive, how they choose to slow down for the landing on the planet, which propulsion system they use, and so forth. In any situation, climbing out of the deeper gravity well will be a major disadvantage.

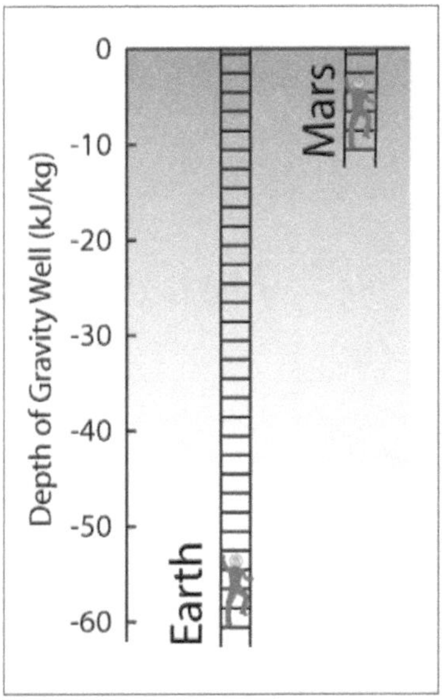

Figure 3.1. Comparison of gravity wells. ***Source*: T. K. Rogers.**

The best way to indicate the "depth" of the gravity well is to calculate the minimum energy required to climb out. As can be seen in figure 3.1, Earth's gravity well is nearly five times deeper than Mars's gravity well. In other words, at a minimum, it takes at least five times as much energy to climb out of Earth's gravity well than to climb out of Mars's gravity well. And when we take into account more factors than just the gravity force, the situation gets even worse.

THE SPACE ELEVATOR

There are no ladders for climbing out of gravity wells, at least not yet. There is a proposed space elevator consisting of a large counterweight, possibly a space station, located more than 22,236 miles above the Earth's surface—the required distance for geosynchronous orbit—attached to Earth's surface by a super strong cable-like ribbon. Elevator cars with more capacity than the retired space shuttle would go up and down the ribbon, transporting passengers and equipment into and back from outer space. Simply reaching geosynchronous height would reduce Earth's gravity by 97 percent, almost the same thing as being completely out of Earth's gravity well. However, the proposed space elevator could go even higher.

Of course, the space elevator has a few challenges: As yet, there is no available material strong enough and light enough for the ribbon, and if there were such a ribbon, it would be susceptible to damage from weather conditions, micro-meteors, space radiation, and who knows what else.

Traveling up the ribbon at 100 mph would get passengers to the geosynchronous level in a little over nine days. So the weight of oxygen, food, and water would be an issue. Going faster, say at Japanese bullet-train speeds (375 mph) would help, but keep in mind that bullet trains travel on maglev (magnetic levitation) tracks designed essentially to eliminate friction. Vehicles designed to climb the space elevator ribbon would need to use friction or some other means of hanging on to the ribbon to keep from sliding backwards, a condition that's not conducive to bullet-train speeds.

The weight of fuel required to drive the elevator car would make the situation far worse. Unlike a normal elevator in a skyscraper, this one would not be pulled up by a separate cable that allows the equipment and energy source for powering the elevator to be placed in more convenient locations. While the cable for a typical elevator is wound up by an electric motor, keep in mind that the electricity powering the motor is likely coming from coal burning in a power plant located many miles away.

A significant part of the space elevator research is looking at ways for transmitting the energy to the car, possibly with laser beams. If the massive weight of onboard fuel could be eliminated, the energy needed to climb the ribbon would be much closer to the minimum energy required for an object to reach geosynchronous height or beyond. In other words, if the fuel does not have to be lifted, then only the mass of the load and container it resides in needs to be lifted.

Unfortunately, until the space elevator gets going, if we're going to send people into space, we're stuck with rocketry, and it has multiple problems created by at least three big issues as described below.

The Downside of Using Rockets to Go into Outer Space

Rockets require a very high velocity in order to reach outer space: Throw a ball upwards, and it will go a short distance before falling back to Earth. Fire it upwards out of a cannon at a much higher velocity, and the ball will go much farther but still eventually fall back to Earth. Put more gunpowder in the cannon, and the ball will go even farther (assuming that the cannon's barrel stays the same length). Keep increasing the amount of gunpowder (assuming the gun and ball don't blow up), and eventually the velocity will be so high that the ball will travel out of the Earth's gravity well and not fall back. This is called reaching escape velocity.

Indeed, nineteenth-century science fiction writers such as Jules Verne in *From the Earth to the Moon* and H. G. Wells in *The War of the Worlds* both depicted space travel using cannons. They evidently judged it to be the most feasible method.

Unfortunately, in order to reach escape velocity the acceleration of the object coming out of the cannon's barrel is a little high for shooting people into space. If the cannon's barrel were a mile long (1.6 km) and the pressure in the barrel constant when fired, the acceleration for reaching escape velocity would be about 4,600 g. Normally an acceleration of 100 g is considered fatal. However, the level for a fatality could be as low as 4 g depending on the person's health, orientation (standing, sitting, etc.), and the length of time the acceleration is experienced.[2] Of course, the acceleration problem could be solved by making the cannon's barrel a good deal longer, but at some point, it would become more challenging than building a space elevator.

Rocketry solves the cannon problem by putting the propellant in a tube behind the payload being sent upwards. This eliminates the need for the lengthy cannon barrel and makes it possible to keep accelerations low enough for human survival. The downside is that the rocket has to use fuel not just for lifting the mass of the payload and rocket but also a substantial amount of extra fuel to lift the mass of its fuel. Fuel is burned and exhausted out the back of the rocket, and so the rocket's mass gets lower and lower as it climbs higher, but it still requires much more fuel to send a payload into space than the minimum theoretical amount.

Then there is the issue of air resistance. High velocity creates high air resistance that in turn creates even more need for fuel to overcome it. Rocketry works for putting payloads into space but isn't cheap. It used to cost about

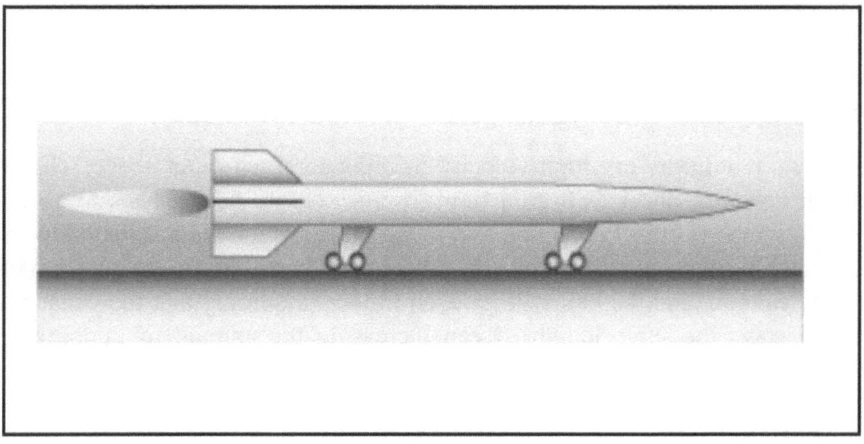

Figure 3.2. Horizontal Rocket. *Source*: T. K. Rogers.

$10,000 per pound just to put a payload into near-Earth orbit using the now decommissioned Space Shuttle or about $1.5 million for a 150-pound person.[3] On this basis sending a payload or person to Mars would be, well, astronomical. (Of course, much of the Space Shuttle expense was due to shuttle maintenance, ground crew, and a plethora of bureaucratic expenses other than fuel.)

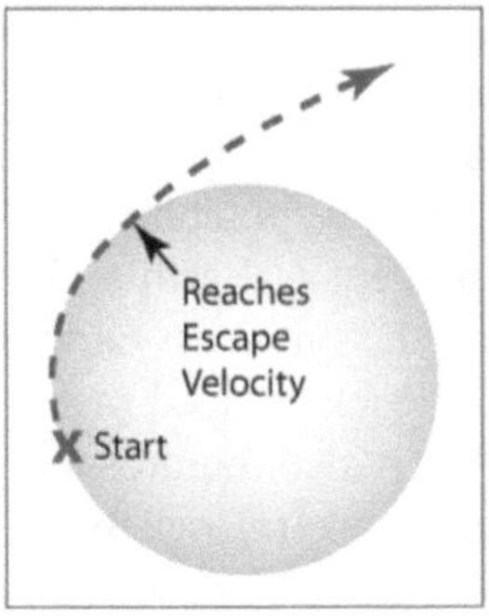

Figure 3.3. Path of rocket launched horizontally. *Source*: T. K. Rogers.

Rockets can't reach the required high velocity instantaneously. The ideal way to launch a rocket into space would be to put it on wheels and launch it horizontally down a very long runway (figures 3.2 and 3.3). When the rocket eventually reached escape velocity, it would appear to lift off the runway and fly into outer space. Wings would not be needed.

While the rocket was accelerating, it would not need to burn extra fuel to counterbalance the force of gravity. This would be done by the wheels. Unfortunately, the runway on Earth would need to be about 1,000 miles long if acceleration were limited to 4 g, not to mention that the heat generated by the air's friction could melt the rocket before it reached the required minimum speed of Mach 33. One can hardly imagine the sonic boom.

The required runway length for reaching escape velocity with a 4 g acceleration on Mars would be only 200 miles, which is almost doable on a lightly populated planet. But while the atmospheric pressure would be far lower, overheating from "air" resistance would still be a problem.

On the other hand, put a wing on the rocket, and the horizontal launch conceivably could be done on Earth, especially if the wing had powerful jet engines. On reaching the maximum altitude the wing could detach and be flown back to Earth. The rocket would fire and accelerate to the required speed for space travel.

While they are rapidly moving forward in a reasonably thick atmosphere, wings can counterbalance an aircraft's weight using very little fuel. By pulling oxygen out of the atmosphere, jet engines can provide thrust with significantly less fuel weight than rockets. By contrast, rockets must carry their oxygen supply, often in a liquid oxygen tank. Virgin Galaxy sent a rocket ship into suborbital space after it was carried to a peak altitude of 51.4 miles (82.7 kilometers) on December 13, 2018, by a specially designed jet aircraft and released.[4] So the system has been used, but the velocity needed for suborbital flight is not even close to escape velocity. A spacecraft based on a winged system may eventually be capable of reaching near-Earth orbit but is unlikely ever to travel all the way to Mars.

Launching a rocket straight up is the worst way to do it from the standpoint of working against gravity, but from a practical standpoint it is the only way to get significant payloads into space from Earth. It probably also would be the only practical way to do so from Mars. Since Mars's gravity is only 38 percent of Earth's, the penalty for launching in a vertical position would be far less than on Earth.

Rockets need mass: It takes a lot of energy to launch a rocket into space, but it also requires a great deal of mass that has to be expelled out the back of the rocket. Rockets don't need to push on anything in order to fly into space. Instead, they work on a concept called conservation of momentum. In other words, the momentum of the rocket going forward equals the momentum of the gases expelled out the back. Since momentum equals mass times velocity, in order for a rocket to go faster, it has to expel more mass out its back side or expel the mass at a higher velocity. Some types of fuel and motor combination can expel their mass at higher velocities than others, which helps minimize the total energy required to go into outer space, but all rockets have to expel mass backwards in order to go forward regardless of fuel type.

The Martian Space Travel Advantage

Reaching escape velocity: The penalty from the mass supply rockets need and from the gravity-well depth can be understood by looking at the ratio of the mass of a fully fueled rocket required for a rocket to reach a given location to the rocket's mass without fuel (the F/E ratio). The mass F/E required to escape Earth's gravity well is 25.6, given a high-performance methane/oxygen fuel. The same ratio for a Mars gravity well would be only 4.4.[5] But keep in mind that for a given payload, the size of an empty Martian rocket would be significantly smaller than an empty Earth rocket. Both the fuel cost and the equipment costs for Mars-based launches would be far less than for Earth-based ones (see figure 3.4).

Figure 3.4. Relative size of rockets required for escape velocity from planet's surface, given identical payloads (payload indicated by the solid color). *Source*: T. K. Rogers.

Reaching the Moon and Ceres: Even though it is a much greater distance and would take a lot more time, the F/E ratio for a journey from Mars to Earth's moon would require only about 21.7 percent of the F/E required to go from Earth to the Moon. On this basis, F/E for a trip to the largest asteroid in the Kuiper belt, Ceres, from Mars's surface would be about 7.3 percent of the F/E for a trip from Earth's surface to Ceres.[6] Supporting asteroid exploring and mining operations would be far less expensive from Mars than from Earth.

Over time Mars would become the space-travel equivalent of a seafaring nation like England. While it would take a while to build up its mining operations, manufacturing capabilities, and space fleet, Mars eventually would surpass Earth in space technology and capabilities. From a military and logistics standpoint, it would be much easier for Mars to invade Earth than the other way around.

From a commercial standpoint, sending goods to Mars would be far more expensive than sending goods back to Earth. Even given this advantage, only goods unique to Mars—genuine Martian rocks, or if any existed, Martian fossils—would be able to bear the cost of transportation to Earth. Raw materials sent back to Earth more likely would be sent from mining and refining operations supported from Mars but located on asteroids. While the cost of launching a payload into space from Mars would be far less than on Earth, it would still be higher than launching the same payload from an asteroid.

The Martian Information-Gathering Advantage

From the high ground of Sun Tzu's day, one could view the battlefield more clearly, a distinct advantage for information gathering. In this regard, the high ground within the solar system would be indicated by the ability of its inhabitants to make astronomical observations.

Earth's ability to gather astronomical information using optical telescopes is limited by the size of its telescope mirrors and the Earth's atmosphere. Really large telescope mirrors currently run in the range of about 10 meters in diameter. Images from multiple telescopes can be combined together using software, which certainly helps with resolution but is not a cure-all. Given the fact that Mars has only 38 percent of the gravity level found on Earth, in theory even larger telescopes could be built on Mars. It would be a long time before Martians had the precision manufacturing capability for producing such equipment.

The latest thing in telescope design, however, uses rotating reflective liquid mirrors that are far easier to build than the glass-type ones. Rotating a pool of a reflective liquid causes its surface to morph from perfectly flat to a perfect parabolic mirror. These telescopes originally used mercury as the liquid but now have other less toxic materials available, including materials designed not to freeze at cryogenic temperatures. Unfortunately, the mirrors are harder to aim because they can't be tilted, but it's not as bad as it sounds since there are ways to optically aim them. On the other hand, the cost of a liquid mirror is about 1 percent of the cost of a comparable glass mirror.[7] Liquid mirror installations are also much quicker and easier to build.

At present liquid mirror telescopes are a little smaller than the large-sized glass mirror telescopes, but there is a proposal for placing a 100-meter diameter liquid mirror telescope on the Moon. This would take advantage of the cold temperatures and the lack of an atmosphere, advantages that enhance image clarity as well as the ability to observe in the infrared and ultraviolet parts of the electromagnetic spectrum.

A terraformed Mars would not have the Moon's super-cold temperature and lack of an atmosphere, but even a terraformed Martian atmosphere would have at most only a third of Earth's atmospheric pressure—and it would contain little water vapor, a component that absorbs infrared. Martian mountains typically are much higher than Earth's and could provide platforms for observatories with even less atmospheric pressure or humidity. Olympus Mons, for example, is three times higher than Mount Everest.

Because of surface winds, Mars would have more dust problems than the Moon, but would still be the better location for observatories. Unlike a lunar outpost, a colonized Mars would have manufacturing facilities and lots of technically trained personnel on hand both to build and maintain the equipment. On the Moon, most of the equipment would have to be launched at absurdly high expense from Earth.

The large telescopes on Mars would have the resolution and light-collecting power to help detect and monitor small-sized and dimly lit asteroids or comets that might endanger Earth. And in the event that one is on a collision course with Earth, the Martian space fleet could very well be better placed for and certainly more experienced in missions to alter the paths of asteroids. In the absolute worst case, if Earth experienced a human-extinction-level disaster, a Mars colony could preserve the human race.

Information, from asteroid monitoring for safety and mining purposes or from pure scientific research, would be a key export from Mars. One need only read a few articles on the Hubble telescope to realize the entertainment, let alone scientific, value it has generated through its thousands of remarkable images. While it's hard to put a dollar value on information, especially its entertainment worth, the data output of a Martian colony could positively blow away a hundred Hubble telescopes.

As for military purposes, a vast and flexible ability to make astronomical observations also would increase a planet's ability to detect and subsequently destroy incoming invaders from outer space. Between Earth and Mars, Mars does indeed hold the high ground in terms of gravity well and physical features conducive to observing and monitoring what goes on in the solar system.

NOTES

1. Sun Tzu, *The Art of War: Everyman's Edition*, translated by Peter Harris (New York: Alfred A. Knopf, 2018), esp. 61–63, 69, 73–74, 167.
2. NASA, "Advanced Space Transportation Program: Paving the Highway to Space," accessed June 5, 2019, https://www.nasa.gov/centers/marshall/news/background/facts/astp.html.
3. Scott Snowden, "Virgin Galactic's Epic 1st Spaceflight Inspires Richard Branson's Message to Grandkids," accessed June 5, 2019, https://www.space.com/42809-virgin-galactic-first-spaceflight-richard-branson-video.html.
4. Robert Zubrin, *The Case for Mars* (New York: Free Press, 1996), 227.
5. Ibid.
6. William Harris, "How Lunar Liquid Mirror Telescopes Work," accessed June 5, 2019, https://science.howstuffworks.com/liquid-mirror-telescope1.htm.
7. NASA, "Liquid Mirror Telescopes on the Moon," accessed June 5, 2019, https://science.nasa.gov/science-news/science-at-nasa/2008/09oct_liquidmirror.

Chapter Four

Space Mining: A Lesson in the Periodic Table of Wealth

While the Spaniards saw colonization as a top-down military-like operation, the English tended to view it more as a commercial venture.[1] Yes, the British government clearly exerted controls, but it also granted a considerable amount of local autonomy to American colonists while chartering commercial ventures such as the Virginia Company with the goal of establishing New World settlements. The English colonists had read about the earlier Spanish efforts and did hope to find precious metals but weren't as obsessed nor as successful with it. Partly by default, Britain ended up focusing on developing trade for agricultural products like tobacco.

Unfortunately, the English colonists did not find cities of gold, and it eventually became clear to the British government that its New World colonies were not a moneymaking enterprise, in part because of the need to maintain troops in America. These troops were necessary for protecting the colonists from conflicts with local Native American tribes and were very expensive—far exceeding the revenues from the new colonies. Eventually, the British government attempted to offset its losses by imposing a tax on tea and other finished goods, a business decision that was not well received by the colonists. Given a money-losing situation, Britain might have fared better to simply walk away. Instead, it chose to stay and chase the sunk costs, only to end up being ousted.

The first Boer War in South Africa started in December of 1880 and followed a similar pattern. After the British had defeated troublesome local tribal groups, a sturdy group largely made up of Dutch farmers, the Boers, decided they had no need for British authority and declared their independence. Britain responded with military force. This led to a series of humiliating British defeats ending in the battle of Majuba. In it, under cover of darkness, the nattily attired British troops—wearing bright colored tunics (most notably red for the infantry) with shiny brass buttons and white pith helmets—quietly sneaked to the summit of a flat-topped mountain overlooking the Boer camp far below, well beyond rifle range. It was a defensive position the British considered impregnable. In the morning, the Boers responded by attacking up the slopes, hiding behind rocks and firing as they went, essentially creating a long-range rifle match. Not only did British attire mark them as targets, but it turned out that long-range rifle shooting was the Boers' national sport. By the end of the day the British had fled in panic with heavy losses, while the Boers suffered almost no casualties. The war ended with Britain capitulating less than a year after the hostilities began.[2]

Even after allowing the Boers to become independent, the British Empire still controlled most of the territory in the southern tip of Africa, territory originally colonized by the Dutch in 1652. In 1795 Britain captured South Africa from the Dutch, only to hand it back in 1803, then retake it yet again in 1806, ownership being finalized by a treaty in 1814.[3] Given the remoteness and the untamed nature of the land and its indigenous population, the back-and-forth conflict, which occurred before the discovery of diamonds and gold deposits, would seem to have been senseless were it not for the fact that South Africa was a valuable control point on the trade route between Europe and the Far East. Compared with losing most of the North American continent following its conflict with the American colonists, the loss of some territory in South Africa was not particularly serious.

Unfortunately for the Boers' independence, they eventually discovered major gold deposits, thereby creating a powerful new incentive for renewed British interference. This time, the British sent massive numbers of well-supplied soldiers against the much smaller Boer army. Gone were the spiffy red tunics, shiny buttons, and white pith helmets. This time, even with yet another early string of disastrous British defeats, gone was any thought of backing down.

Eventually, the British overwhelmed the Boer army and occupied the Boers' territory. While surrender would have been the rational choice, by now the Boers themselves had invested way too much blood and treasure to do so. Hence, the conflict devolved into a nasty guerrilla war. In order to deprive the Boer fighters of a support system, the British burned down thousands of farms and forced the displaced Boer women and children into prison camps euphemistically called "concentration camps."

Crowded into poorly run and hastily set up camps with poor sanitation, many died of disease, especially children. Faced with such a ruthless response that denied them any reliable means of supply for even basics like food and clothing, the Boer fighters eventually surrendered.

History shows that when hemorrhaging wealth, entities with a sense of entitlement will double down even when it would be better to walk away.[4]

There's something about having invested substantial resources that makes it incredibly hard to give up. In the business world the phenomenon is called the fallacy of sunk costs. However, let the enterprise become profitable, and the entity that considers itself the rightful owner will self-righteously abandon even common decency in order to maintain or gain control. There is no reason to expect that space mining would be an exception.

ASTEROID PROFIT PROJECTIONS VERSUS COST REALITIES

Space mining offers virtually unlimited access to a periodic table of materials in concentrated form. It is a dream almost as good as being able to perform alchemy. A single asteroid can contain trillions of dollars' worth of material floating in space, presumably "free" for the taking. For example, the asteroid scheduled to be explored by NASA in 2026, 16 Psyche—a gigantic chunk of metal 213 kilometers in diameter—is hyperbolically estimated to have a value of $\$10^{19}$.[5] Divide that worth by about 10 billion—a rough estimate of the number of people in the world by the time asteroid mining becomes a reality—and it comes out to about a billion dollars for each person on Earth. Not bad, but also not achievable considering that such a large influx of material would dramatically drive down its market value. Then there are also the considerable costs of mining and transportation (see the Spanish example in chapter 1).

NASA's OSIRIS-Rx mission launched in 2016 will be the first to capture and return to Earth a sample from a near-Earth asteroid, at a cost of nearly a billion dollars. NASA's Psyche mission will cost a similar amount and not even return a sample. The spacecraft simply will orbit the asteroid while photographing and analyzing it from a distance using an array of instruments. Of course, private industry should be able to beat NASA's costs substantially by launching numerous small low-cost probes entirely focused on finding minable asteroids rather than conducting scientific research like a NASA project. Nonetheless, even the asteroid exploration phase is going to require an investment in the billions of dollars.

Then there is also the investment in time required for exploring asteroids. The OSIRIS-Rx will require about seven years to return to Earth a relatively small sample of the asteroid it visits.[6] Imagine prospecting for gold in California starting in 1849 by analyzing a sample every seven years at a cost for the time period comparable to a billion dollars in today's money. Fortunately, at least some of the asteroid prospecting can be done with telescopes. Furthermore, properly configured probes sent to actual asteroids can perform a spectacular array of analysis without sending back a sample to Earth. Nonetheless, there is a real need for improvement both in terms of cost and in terms of time required—improvement that likely will happen, at least to some extent.

Actually mining an asteroid and returning its valuable parts to Earth will require tens of billions of dollars in additional investment over the exploration phase, much of which will be needed to fund the creation of equipment that does not yet exist—a perfect setup for creating excessive amounts of sunk cost. Movies normally depict astronauts providing the onsite intelligence needed for asteroid mining ventures, but given the difficulty of keeping astronauts alive and healthy, such is not likely to happen. Remote-control operation from Earth also is not practical due to communication time lags that could be the best part of an hour between sending a control signal and receiving visual feedback on its result. Most mining operations will need to use sophisticated robotic devices controlled largely by onsite artificial intelligence—again, technology that does yet exist.

The low mass and subsequent low gravity available on an asteroid will pose still another problem. Setting off a blast on the asteroid's surface, a common mining practice on Earth, could significantly alter an asteroid's position or possibly its rotation. Blasting off from the asteroid's surface with a payload could have a similar effect. The asteroid might need to be outfitted with thrusters to keep its movement under control. It might actually be desirable to convert the asteroid into a type of space transport and move it to an orbit closer to Earth or Mars.

Escape velocity for leaving the asteroid might be so low that it could pose a danger to mobile equipment traveling on the asteroid's surface. If the equipment is not secured to the surface and moves too fast or bounces too much, it could end up leaving the asteroid's surface and be lost in space. The mining operation would require considerable amounts of power, which in turn might require substantial arrays of solar cells or even a small-sized nuclear reactor. Given all the costs of exploiting an asteroid, ownership is going to be a big deal.

The Issue of Ownership

Unfortunately for asteroid ownership claims, resolutions adopted by the General Assembly of the United Nations in the 1960s basically state that outer space and the various celestial bodies in it are owned by all of humanity.[7] In other words, being the first to occupy or mine an asteroid does not count for ownership. Still, it's pretty ridiculous to think that asteroid mining entities are going to sit around a campfire and sing "Kumbaya" while distributing their profits to all of humanity. They are going to expect a return on the billions they have invested and will have a powerful incentive to ignore the niceties of international law.

Given this situation, the U.S. Congress rode to the rescue (or totally mucked things up, depending on one's perspective) by passing the Space Act of 2015.[8] This essentially grants the mineral rights for asteroids or other celestial bodies to any U.S. company that sets up a mining operation on them. While this opens the door for commercial space mining operations, it also opens the door for competition and possible conflict.

Assuming the ownership hurdle has been overcome and asteroid mining begun, the cost of transporting a pound of product back to Earth will be expensive but significantly less expensive than launching a pound of material into low Earth orbit (in 2016 around $800 per pound using the Falcon Heavy, but according to SpaceX, it eventually could be under $100 per pound).[9] Many factors could make a difference such as using rocket fuel produced on an asteroid instead of launching it from Earth and by taking advantage of air resistance rather than using rocket fuel for slowing down once the payload's container reaches Earth's atmosphere. This has been done on other space vehicles with parachutes and with glider-like designs, such as NASA's Space Shuttle, which took advantage not just of aerodynamic drag but also lift.

Launching cargo containers from Earth and transporting them to the asteroid to be filled with product prior to a return trip also would be a significant cost. This possibly could be reduced by fabricating cargo containers on Mars and by taking advantage of Mars's much lower gravity well when launching them into space. After reaching an asteroid mine, the containers would be filled with product and be sent on a one-way trip to Earth. Better yet would be the possibility of manufacturing the containers on the asteroid itself.

Given the costs of transport, the candidates for asteroid mining output would need to be specialized high-value, low-volume products if they were to be transported for consumption on Earth. Commodity metals such as iron, aluminum, and copper would not be good candidates because the market size on Earth is far too large to be supplied with a limited number of spacecraft. They also are readily available on Earth, a fact that drives down their per-pound value.

At this point, it's nearly impossible to calculate asteroid mining costs. Nevertheless, it's reasonable to assume that any material worth around $1,000 a pound or more on Earth would be profitable, even in the early years of asteroid mining. Gold and platinum group metals—platinum, ruthenium, rhodium, palladium, osmium, and iridium—typically exceed this value by a factor of about 10.

Platinum group metals have numerous industrial uses, such as automobile catalytic converters and catalysts for everything from chemical-making processes to fuel cells. Platinum group metals don't naturally occur on the surface of Earth. Deposits of them are typically associated with ancient meteor impacts; they actually may have been transported to Earth on the meteors.[10] Certainly, the platinum metal content of meteors is higher than on Earth's surface. These factors alone imply that mining platinum group metals on asteroids could be viable.

As the cost of mining improves, rare earth metals would constitute another asteroid mining possibility. These metals are vital to a wide range of products from computers, smartphones, and solar cells to the powerful magnets used in wind turbines. The processes required for extracting rare earth metals, however, are difficult and plagued by serious pollution problems. To make matters worse, even commercially viable deposits tend to have low concentrations.

Currently, 85 to 95 percent of the world's supply of rare earth metal supply comes from China, which has strategically used its rare earth metals to create significant economic leverage.[11] Although China has large deposits of them, they are only part of China's advantage. China has acquired the infrastructure and know-how required to extract them. Certainly, China's cheap labor also has helped it become the low-cost behemoth in rare earth metals, but the biggest factor in China's success may be its lax environmental laws.

Countries importing rare earth metals from China are in a sense exporting the environmental problems that would be experienced with domestic production, and this may be yet another reason to obtain the metals from asteroids. From a biological perspective, asteroids don't have an environment to destroy. If most of the otherwise nasty processing required to obtain them were done in outer space, rare earth metals would be a much greener product for use by Earthlings.

As for profitability, if an asteroid mine's transport container is only partly filled with gold and platinum group metals, filling the rest of its capacity with less valuable rare earth metals might improve profitability over sending back a partly empty container. The only additional expense would be the extra fuel for sending the pounds of rare earth metal from the asteroid to Earth. The higher capital cost of the transport container and other associated transport expenses already would be covered by the gold and platinum group metals.

Sending materials to Earth is not the only moneymaking possibility. Ordinary substances like iron along with its alloy steel also may end up being significant moneymakers. While the price on Earth is pennies per pound, sending a pound of steel just into near-Earth orbit currently costs around $800 using the lowest priced transport system. Even if the cost of transportation from Earth drops to $100 per pound, sending a pound from an asteroid mining/manufacturing operation to another location in outer space would still cost less. Given the strength of steel, cheap availability of it would open up all kinds of possibilities for space-based structures. What's more, M-type asteroids are mostly made of iron and nickel that would likely need a minimal amount of processing to be useful. For that matter, any useful raw materials mined in outer space that remain and are used in outer space typically will be worth significantly more than their value on Earth.

C-type asteroids, for example, contain carbon compounds and also can contain around 20 percent water.[12] Due to their low albedo, they can be hard to detect and are not generally considered prime candidates for mining rare and precious metals to be sold on Earth, but their water content alone would be a highly valuable commodity for use in outer space. At current rates, launched from the surface of Earth just to low Earth orbit, a liter of water would cost around $1,800. Water from an asteroid presumably would be far cheaper. Of course, there are the mundane uses for water, such as drinking or growing food using hydroponics. Indeed, the carbon in c-type asteroids as well as the availability of phosphorus in them are both needed for food production. Water also could be broken into hydrogen and oxygen using electrical power from solar panels. The oxygen could be used for breathing or the oxygen and hydrogen could be stored in separate tanks for later use as rocket fuel, again forgoing the ultra-high price of having to launch it into space from the Earth's surface. Hence, a spacecraft launched from Earth and headed for Mars would only need to carry enough fuel to reach near-Earth orbit, where it would be refueled for the rest of the journey using fuel produced in outer space.

Of course, not all spacecraft are going to be designed to operate using an oxygen/hydrogen fuel. SpaceX, for example is focusing on using an oxygen/methane mix. Presumably, on a c-type asteroid, with a little more chemistry, the oxygen/methane also could be produced.

S-type or silaceous asteroids are mostly composed of stony materials and nickel-iron.[13] They are thought to have a diversity of metals available in them that could include nickel, gold, and platinum. The presence of silicon suggests that S-type asteroids also could provide silicon for creating the type of large, flawless crystals that could be grown in microgravity conditions inside growth chambers orbiting the asteroids being mined. Those crystals could be used for everything from semiconductor chips to solar cells.

Of course, asteroid mining does not require the colonization of Mars, but it would benefit from it. The journey to Mars undertaken by thousands of colonists would help provide an outer space market for water, fuel, and other materials produced from asteroids, a market that would help finance asteroid mining. Subsequently, mining entities are likely to be strong advocates for Mars colonization. Given its much lower gravity well and superior conditions for astronomy, Mars would be better than Earth as a location for exploring asteroids and for setting up and maintaining the mining/manufacturing equipment on them.

The Interaction of the Mars Colony with Asteroid Mining

Mars, like South Africa, also could end up not just as a valuable control point in asteroid mining, but also a source of mineral wealth itself. While Mars is a much smaller planet than Earth, it has about the same amount of land area as Earth.[14] Furthermore, at this point almost none of Mars has been explored by anything actually on its surface. By comparison, imagine the chances of finding gold on Earth if it only had been explored by a handful of robots driving a few dozen miles around Earth's surface. Finding gold or other platinum group metals on Mars would require a lot of exploration. If a significant deposit were found, it might be easier to mine and refine it on Mars and then ship it to Earth than to do the same process on an asteroid.

There are also the 49 strategic metals, a group that includes the rare earth metals, which could end up being mined on Mars or on asteroids.[15] These metals are essentially irreplaceable in applications orders of magnitude more valuable than the strategic metals in them. While the metals themselves may not be all that rare on Earth, finding a high enough concentration of them to be economically mined can be problematic. The total annual production of these strategic metals added together is typically less than the total production of copper. In fact, the total annual production of some strategic metals could fit in a few boxcars.[16] A single BFR spacecraft returning from Mars with a 50-ton load of one strategic metal could represent a significant share of its market on Earth.

Using the SpaceX transportation model, BFR spacecraft serving Mars have to be returned to Earth and repeatedly reused in order to keep passenger transport costs low. Hence, loading the spacecraft with product such as precious metals, platinum group metals, or strategic metals makes good economic sense. Given a BFR's cargo capacity from Mars of 100,000 pounds, if loaded with platinum—an admittedly best-case scenario—the cargo would be worth around $1.7 billion. Compare this to the revenue of around $25 million for transporting 100 people to Mars (at a ticket price of $250,000 per person), and the return trip could be the greatest profit maker. Unlike a cargo of people, a cargo of metals does not require additional food, water, or oxygen. It only requires a somewhat modest extra fuel cost to be launched from Mars on a return trip to Earth.

In addition, products directly traceable to Mars, from gemstones to fossils (assuming that Mars once contained life forms), could command exorbitant prices if transported to Earth. These items do not need to have any utilitarian or rational value. South African diamonds are an analogous example. The De Beers Company has not just run South Africa's diamond mines but also carefully controlled the supply to prevent a declining price due to oversupply. By manipulating popular culture—"diamonds are a girl's best friend," "diamonds are forever," "an engagement ring must be a diamond costing three months' salary," and so forth—De Beers has created a multibillion-dollar industry that sells small chunks of carbon, albeit sparkly chunks, for exorbitant prices.[17]

Of course, there are downsides to filling the transport spacecraft with Martian products bound for Earth. First, if the products are strategic metals, there is the significant time delay between extracting the metals and being able to market them on Earth. This could create a boom-or-bust supply situation that would need careful management. Second are the political problems that might invite unwanted interference in controlling the supply.

Any disruption in the supply of critical metals could have serious consequences on Earth. For example, smartphones use 70 of the 87 nonradioactive, naturally occurring elements from the periodic table including 16 of the 17 rare earth metals, and although an individual phone has only a tiny amount of rare earth metal in it, the phones—comprising a trillion-plus dollar a year industry—cannot be made without them.[18] This type of potential economic impact invites unwanted meddling. It also ensures that the Earthling groups that have invested in creating the Martian supply would not walk away without a struggle.

Investors will not be the only group directly involved in exploiting extraterrestrial mineral deposits. Someone or something will need to venture into outer space to provide the intelligence needed for exploitation of these resources. If this need is met by sending people, keeping them alive, healthy, and content will vastly complicate the endeavor.

NOTES

1. Niall Ferguson, *Civilization: The West and the Rest* (New York: Penguin Books, 2011).
2. Byron Farwell, *The Great Boer War* (Barnsley, South Yorkshire: Pen and Sword Military, 2009).
3. Ibid.
4. Jack Snyder, *Myths of Empire: Domestic Politics and International Ambition* (Ithaca, NY: Cornell University Press, 1991).
5. Karla Lant, "NASA Is Fast-Tracking Plans to Explore a Metal Asteroid Worth $10,000 Quadrillion," accessed June 5, 2019, https://futurism.com/nasa-fast-tracking-plans-explore-metal-asteroid-worth-10000-quadrillion/.

6. Nola Taylor Redd, "OSIRIS-Rex: Bringing Home Pieces of an Asteroid," accessed June 5, 2019, https://www.space.com/33776-osiris-rex.html.

7. United Nations Office for Outer Space Affairs, "Treaty on Principles Governing the Activities of States in the Exploration and Use of Outer Space, including the Moon and Other Celestial Bodies," accessed June 5, 2019, http://www.unoosa.org/oosa/en/ourwork/spacelaw/treaties/outerspacetreaty.html.

8. Stephen D. Krasner, *Sovereignty; Organized Hypocrisy* (Princeton, NJ: Princeton University Press, 1999).

9. Richard Yonck, "The Dawn of Space Age Mining," accessed June 5, 2019, https://blogs.scientificamerican.com/guest-blog/the-dawn-of-the-space-mining-age/.

10. Peter B. de Selding, "SpaceX's New Price Chart Illustrates Performance Cost of Reusability," accessed June 5, 2019, https://spacenews.com/spacexs-new-price-chart-illustrates-performance-cost-of-reusability/.

11. David Tilley, "Platinum Group Metals," accessed June 5, 2019, https://www.geologyforinvestors.com/platinum-group-metals/.

12. Chris Lo, "The False Monopoly: China and the Rare Earths Trade," accessed June 5, 2019, https://www.mining-technology.com/features/featurethe-false-monopoly-china-and-the-rare-earths-trade-4646712/.

13. Stephen Shaw, "Posts Tagged 'C-type Asteroids' Asteroid Mining," accessed June 5, 2019, http://www.astronomysource.com/tag/c-type-asteroids/.

14. Nancy Atkinson, "What Are Asteroids Made of?," accessed June 5, 2019, https://www.universetoday.com/37425/what-are-asteroids-made-of/.

15. Tim Sharp, "How Big Is Mars?," accessed June 5, 2019, https://www.space.com/16871-how-big-is-mars.html.

16. David S. Abraham, *The Elements of Power: Gadgets, Guns, and the Struggle for a Sustainable Future in the Rare Metal Age* (New Haven, CT: Yale University Press, 2015).

17. Eric Goldschein, "The Incredible Story of How De Beers Created and Lost the Most Powerful Monopoly Ever," accessed June 5, 2019, https://www.businessinsider.com/history-of-de-beers-2011-12.

18. David Nield," Our Smartphone Addiction Is Costing the Earth," accessed June 5, 2019, https://www.techradar.com/news/phone-and-communications/mobile-phones/our-smartphone-addiction-is-costing-the-earth-1299378.

Chapter Five

Humans or Humanoids: Why People Are a Bargain

By the time Mars is ready for colonization, multipurpose humanoid robots (H-bots) will be commonplace on Earth. These will not be the fixed-position robotic arms found in manufacturing facilities or various types of self-driven cars or tractors. Even though these may have a plethora of sensors, a considerable amount of artificial intelligence (AI), and the ability to communicate verbally with humans, they are not H-bots. H-bots will have a human-like appearance, flexibility, and adaptability. They will be general-purpose devices as opposed to those designed for specific tasks.

By definition, H-Bots will have the following characteristics:

1. Two legs (rather than just using wheels or tracks) that provide upright mobility in just about any terrain and that let the H-bot change height by kneeling, sitting, or stretching upward similarly to how a human would behave.
2. One or two arms with multipurpose grippers or hands at the end that can perform similar functions as human arms and hands. Of course, H-bots could have multiple arms or tentacles, but this would achieve weirdness with, in most cases, only marginal extra functionality. The upright position of the H-bot would allow the arms to be used while the H-bot was moving.
3. A head-like structure that rotates, tilts, and contains sensors including one or two video cameras (robot eyes). More than two eyes could be used but would make the bot look spiderlike. Two eyes, though, are needed for depth perception.
4. A torso to which the above structures are attached. The torso most likely would contain the support equipment for powering the H-bot and probably contain some distributed processing capability (similar to networked computers) dedicated to controlling specific devices like arms, legs, and touch

sensors. In higher-level biological creatures this distributed processing is contained in places like the spine and intestinal tract. Yes, weird as it is, in humans, the intestinal tract is loaded with neurons in a brain-like structure that controls the somewhat complicated chemistry of digestion.

5. A powerful general-purpose computer (brain) with enough AI and machine-learning capability to act at least somewhat autonomously and to communicate with humans. In higher-level biological creatures these general-purpose brains are placed in the head so they have close proximity to major sensors—eyes, nose, ears—that require a great deal of processing power and are key to rapid decision making. H-bots probably would follow a similar pattern.

While humans are not the strongest, fastest, most perceptive, or most agile creatures on Earth, they have by far the most adaptable design—primarily due to hands attached to arms that are not needed for mobility—backed by the highest level of general-purpose intelligence. So, it's reasonable that general-purpose machines designed to serve people would have some similar design features. In fact, it's hard to beat a biological design resulting from millions of years of evolution.

The consumer version of H-bots likely will cost about as much as a compact car, up to the price of a luxury model depending on the list of features. These H-bots will handle routine household chores such as cleaning, taking out the trash, changing diapers, mowing lawns, and walking dogs (yes, people will still have dogs and other pets). And while we humans think of these as mundane tasks requiring little skill or intelligent thought, any one of them actually is incredibly hard to automate. Building a single compact, self-contained, mobile device capable of doing all the tasks and of learning new ones is quite a challenge.

The more expensive and capable versions of H-bots would find their way into various service jobs such as personal assistant, janitor, and so forth. Once again, while the activities of these jobs may seem straightforward to humans, they are actually challenging to implement with technology and would benefit from the availability of a highly developed general-purpose design. However, the most basic requirement of all H-bots will be their ability to work and interact directly with a wide variety of humans.

Unfortunately, this requirement of interacting with humans poses a considerable design dilemma. It's called the uncanny valley.[1] A robot tends to elicit feelings of familiarity or acceptance as it approaches a humanlike appearance, that is, until it gets too close but not quite there. At this point acceptance by real humans crashes into feelings of eeriness or revulsion. Motion intensifies the emotional response of humans to an H-bot more than what it would be if the H-bot were stationary. Initially, this helps, but at the bottom of the uncanny valley, motion creates increased revulsion. The form of a human corpse represents the low point of the stationary version for the uncanny valley curve, which is bad enough, but animate the corpse and it becomes a zombie—an object of terror.

While humans can learn to overcome their emotions, the bad news is that the uncanny valley is probably hardwired into the human brain, possibly as a protective mechanism for preventing the spread of disease. If a person's appearance is not quite right, it could indicate a contagious condition. Certainly, a corpse lying beside the road could be a source of severe contagion.

H-BOT DESIGN OPTIONS

Since human interaction is required of H-bots, designers have three options: industrial, cutesy, or the most difficult and expensive of all, human replica. But, keep in mind that this is more of a spectrum than three distinct design types.

Industrial H-bots that emphasize function with limited resemblance to humans produce a ho-hum response in people. However, give the industrial H-bot a few cute behaviors, and people can develop a certain attachment to it, such as with R2-D2 of *Star Wars* fame. Keep in mind that the actual R2-D2 in the movie was a midget inside an R2-D2 suit. Why a midget instead of a remote-controlled device? The humanlike cutesy behavior of R2-D2 would have been very difficult to implement using an actual machine. Still, even without the cute behaviors, no one is going to be creeped out by a trashcan driving around on roller-skate feet attached to stubby legs while making squeaky sounds and occasionally extending a robotic arm.

Cutesy H-bots are the next step upward toward human acceptance, especially if they have large wide-looking eyes and a neutral or happy-looking mouth and cutesy or slightly wacky behaviors such as C-3PO had in *Star Wars* movies. Like cars, cutesy H-bots would come in different colors and feature packages, but it would not be necessary to make them one-of-a-kind unique in order to satisfy their human owners. Data, the android depicted in the *Star Trek* franchise probably represents the pinnacle of cutesy-type (well, at least somewhat amusing) H-bots located at the peak of the uncanny valley curve that occurs right before the decline to the low point. Data is clearly not human, but has just the right approximation of human appearance and behavior.

Human replicas are the ultimate in H-bot design but also the most dangerous form. They are frequently found in science fiction movies, such as *Blade Runner*. In this depiction, they're called replicants; and when they go wrong, they go homicidally wrong, hence, are hunted and destroyed without hesitation by the movie's main character. But, when they go right, they go very right as depicted by the replicant the main character falls in love with and subsequently spares. Unlike cutesy H-bots, replicas would have to be custom models with the exception of those that resemble famous people. One can only imagine the rental possibilities for parties: a George Washington model for the Fourth of July, a Babe Ruth version for World Series watching parties, or perhaps a Marilyn Monroe type for, well, ah, singing "Happy Birthday."

The Pros and Cons of H-bots as Residents of Mars

Since H-bots can be turned off and require neither food nor water, let alone oxygen during the trip to Mars or for that matter when they arrive, H-bots will be contenders for the colonization of the red planet. True, they will require a power source on the planet and some downtime for charging (or refueling if they use fuel cells) in order to operate, but they will otherwise need no sleep or motivation. At first glance, it might appear that H-bots could pretty much be the only inhabitants of Mars.

However, H-bots designed for Mars's harsh conditions will be much more expensive than even the highest-end consumer models. They will not need to be human replicas but will need to have the mobility over unpredictable terrain and the flexibility and adaptability of human workers. Due to the remoteness of Mars, the resident H-bots will have a far greater need for self-repair than any Earth-based model. Given the limitations of communication with Earth (easily 40 minutes or longer for a brief two-way exchange), the H-bots will need additional autonomy and decision-making authority or, in other words, a much more sophisticated form of AI. They will need to be much more robust than Earth-based H-bots and be capable of resisting extreme temperature variations and dust exposure. To avoid the prohibitive cost of transport, H-bots ideally would need to be able to remote replicate themselves. But in the early years of colonization, the industrial base and raw materials for such capabilities would not be available on Mars.

There also are serious challenges about powering H-bots. At present, battery technology is woefully bad in terms of energy density, either as a measure of stored energy per unit of mass or per unit of volume (see table 5.1). A battery-operated H-bot would be either heavy and bulky or would quickly run out of power when given a task requiring physical work. Fuel cells are better and, thanks to their favorable energy density ratios, were originally developed for powering satellite communication and control systems. But the power needs of communication and control systems are minor compared

Table 5.1. Energy Density of Various Energy Storage Options

Type	*Energy/Mass (MJ/kg)*	*Energy/Volume (MJ/L)*
Plutonium 238	1,600,000	NA
Gasoline	44	32
Fat (animal, vegetable)	37	34
Hydrogen	120	8 (liquid H_2)
Lithium-Ion Battery	0.36-0.95	0.9-2.4
Lithium-Ion Battery Pack (Chevy VOLT)	0.3	0.4

Author-generated from the following sources:

1. Steven D. Howe, Robert C. O'Brien, Troy M. Howe, Carl Stoots, "Compact, Low Specific-Mass Electrical Power Supply for Hostile Environments," Idaho National Laboratory, 2019, http://polarpower.org/PTC/2013_pdf/PTC_2013_Howe.pdf.
2. Office of Energy Efficiency and Renewable Energy, 2019, https://www.energy.gov/eere/fuelcells/hydrogen-storage-basics-0.
3. Ping Zhang, "Energy Density of Fats," *The Physics Fact Book, 2004*, https://hypertextbook.com/facts/2004/PingZhang.shtml.
4. "Lithium Ion Battery," Clean Energy Institute, University of Washington, 2019, https://www.cei.washington.edu/education/science-of-solar/battery-technology/.
5. Fred Schlachter, "Has the Battery Bubble Burst?," APS News, August/September 2012, https://www.aps.org/publications/apsnews/201208/backpage.cfm.

with the power needed to do the type of physical work that might be required of an H-bot.

The size and weight of a fuel cell system in an H-bot again would depend on its fuel. Hydrogen is a likely choice, but even pressurized to 10,200 psi (700 bar) it will still require over six times as much volume to store the same amount of energy as a gallon of gasoline. Indeed, gasoline would be ideal for fuel cells, and small-scale gasoline fuel cells have even been built. An electric automobile using this technology could get well over 100 miles per gallon. Given the massive market for fuel-efficient vehicles, it is likely that large-scale fuel cells of this type eventually will be developed. Of course, on Mars, there is no known source of crude oil that can be used for making gasoline. The situation is further complicated by the fact that fuel cells need an oxidizer such as—surprise, surprise—oxygen, although hydrogen peroxide and halogens are also possibilities. Unfortunately, the widespread use of fuel cells on Mars is not straightforward.

Of course, various types of nuclear fuels would be the ultimate source for H-bots. For example, plutonium used in nuclear decay mode has over 36,000 times more energy per unit of mass than gasoline. What's more, energy storage devices of this type already have been used to power satellites in outer space. The problem is both the available supply, cost, and radiation danger of plutonium. Could plutonium be feasible as a means of powering Martian H-bots? Maybe.

Movie depictions of various H-bot types generally overestimate their future abilities by simultaneously depicting them with superhuman strength and athletic ability. However, given the various problems with powering them while keeping their weight low, H-bots will be doing well just to match human capabilities in these areas. The considerable hype concerning exoskeletons under development by the U.S. military may seem to indicate otherwise. These are wearable external skeleton-type devices that are battery-powered and designed to increase the ability of soldiers to carry a greater weight of equipment and supplies over greater distances.[2] And if an exoskeleton can do wonders for a human, why not design it into an H-bot and create a superstrong one? Unfortunately, weight, bulk, and battery life have proven to pose serious limitations for exoskeletons and although still holding promise, the eventual performance in reality is not likely to match the original dream. Besides, what would prevent humans or even H-bots from putting on exoskeletons as needed? Great strength is not needed for performing many, if not most, of the tasks H-bots might be expected to perform—cleaning toilets, diaper changing, dusting, painting, welding, wiring, weeding, and so on.

The Pros and Cons of Humans on Mars

In contrast to H-bots, an adult human who costs his parents—not the developers of Mars—over $260,000 (in 2017 dollars) just for the basics to raise may seem like a bargain.[3] Include an average cost of a K–12 education in the United States of about $160,000 (2013 to 2014) and the total spending per adult rises to $420,000.[4] Add the average cost of a four-year public college degree at $177,000 (2015 to 2016), and the total becomes $597,000.[5] This is a bare-bones estimate that does not include possibilities such as extracurricular training, private versus public schools, or advanced degrees.

In addition, unlike H-bots, human immigrants will be a source of revenue, willing to pay several hundred thousand dollars for a one-way ticket to the red planet. They also will come with a ready-made capability for self-repair since they not only have immune systems but also a built-in ability to heal injuries. As for producing additional workers (offspring), given a livable habitat, they definitely would not need a well-developed industrial base for reproduction. And it's worth noting that the energy storage system for humans, fat, has energy densities that are almost as good as gasoline.

Even if super strong H-bots existed, their strength would not be as useful on Mars as it might be on Earth. In the lower gravity environment of Mars, humans would be able to lift about 2.6 times as much mass as they can on Earth. This means that relatively strong people who could lift 100 kg of mass with a weight of 220 pounds on Earth would be able to lift 260 kg of mass on

Mars that would otherwise weigh 572 pounds if located on Earth. Even if an H-bot theoretically could lift significantly more mass than a human, in many cases the larger mass (depending on its density) would be too bulky for the H-bot to grip or to balance without tipping over.

The Security and Emotional Advantages of Humans

By the time Mars is not just being explored but seriously colonized, the novelty of H-bots will have worn off and been replaced in some cases by suspicion, resentment, and disrespect. Displaced human workers will be outright hostile to them as job killers. Unlike the case with the job-killing automation found in factories, humans are going to have face-to-face encounters with H-bots, and it's going to be an uneasy situation.

Earth-based H-bots may have the unintended consequence of precipitating a new wave of vandalism and crime. Armed robbers could take a certain delight in blowing away the H-bots in the businesses they rob. Why not? Killing an H-bot is only a property crime, not homicide. Likewise, smashing an insipid-looking H-bot probably will be an emotional outlet for restless young vandals. For more profit-minded criminals, H-bot theft and an emerging black market for parts and the stolen H-bots themselves might appear. Criminals and vandals even may view themselves as heroic Robin Hoods or as revolutionaries resisting the evils of the machine.

History provides examples of serious rage directed against job-killing machines—for example, the Luddites. These were a group of skilled British textile workers who were replaced by low-skilled, poorly paid workers running automated textile-producing machines. The Luddite movement began in earnest in 1811 when members began smashing textile machines and burning down the houses of the owners. By 1812 the situation had gotten so out of hand that an outraged Parliament made textile machine smashing a capital offense and hanged a few offenders to make a statement.[6] Needless to say, the movement soon died, along with the Luddites unlucky enough to be prosecuted. No doubt, new laws also will have to be passed to protect H-bots.

Eventually, an H-bot will commit homicide followed by major worldwide news coverage dissected in minute detail by analysts and talking heads. As usual, fear of such events will be inflated way beyond rational levels by the massive news coverage. The same thing has proven to be true in the United States with airplane crashes and terrorist attacks. They are both rare causes of death but have generated a disproportionate level of fear and subsequent government spending for prevention.

This H-bot–caused homicide could result from many possible causes. The H-bot involved could be deliberately programmed to kill by a maniacal human, or it could kill as a result of a malfunction or programming error. Since H-bots likely will come with networking capability, they also will be susceptible to hacking. The worst case would be if the homicidal H-bot acted autonomously—in other words, decided by itself to kill a human. No matter what the cause, the first homicide by an H-bot will set off a plethora of conspiracy theories, suspicions, and hysteria.

However unlikely, the fact that an army of networked H-bots could be surreptitiously hacked by a psychopath or foreign power and reprogrammed to violently attack their human masters would be gasoline thrown on the fire of hysteria. Even more terrifying is the thought that an army of networked H-bots with highly developed AI could decide on their own to attack in a coordinated manner.

Imagine the nightmare of dealing with security issues if an H-bot eventually is wired up as a substitute suicide bomber. Certainly, metal detectors will not be helpful for singling out weaponized H-bots. Skeletal parts of H-bots could be made of carbon fiber–reinforced plastic, but it would be hard to completely eliminate metals for wiring, motors, and actuators. Radio-actuated kill switches could be wired into their power supply so that H-bots could be remotely deactivated by security personnel in an emergency, but the switches themselves could be hacked and disabled.

Already, a large number of scientists and engineers, including well-known names such as Stephen Hawking, Elon Musk, and Bill Gates, have begun issuing warnings about the dangers of AI ending human life on Earth.[7] It appears that a situation such as the one depicted in the 1984 sci-fi movie classic *The Terminator*, in which machines with AI take over the world from humans, is an actual possibility.[8] It would not be the first time a new technology has had serious unforeseen or unintended consequences.

In the end, no Mars developer, financial backer, politician, or strategic thinker will want autonomous H-bots—with the ability to learn, remote replicate, and possibly evolve—to have an entire planet to populate and control. While H-bots in their various forms will be sent to Mars as a labor force, they will be strictly controlled by Mars-based humans who will outnumber them.

NOTES

1. Masahiro Mori, "The Uncanny Valley: The Original Essay," accessed June 5, 2019, https://spectrum.ieee.org/automaton/robotics/humanoids/the-uncanny-valley.

2. Sydney J. Freedberg Jr., "Lockheed, Army to Test Exoskeleton in December," accessed June 5, 2019, https://breakingdefense.com/2018/05/lockheed-army-to-test-exoskeleton-in-december/.

3. Alex Glen, "Cost of Raising a Child Tops $260,000—Just for Basics," accessed June 5, 2019, https://www.nerdwallet.com/blog/insurance/cost-of-raising-a-child/.

4. National Center for Education Statistics, accessed June 5, 2019, https://nces.ed.gov/fastfacts/display.asp?id=66.

5. National Center for Education Statistics, accessed June 5, 2019, https://nces.ed.gov/fastfacts/display.asp?id=76.

6. Clive Thompson, "When Robots Take All of Our Jobs, Remember the Luddites," accessed June 5, 2019, https://www.smithsonianmag.com/innovation/when-robots-take-jobs-remember-luddites-180961423/.

7. Michael Sainato, "Stephen Hawking, Elon Musk, and Bill Gates Warn about Artificial Intelligence," accessed June 5, 2019, http://observer.com/2015/08/stephen-hawking-elon-musk-and-bill-gates-warn-about-artificial-intelligence/.

8. Charli Carpenter, "Rethinking the Political Science/Fiction Nexus: The Campaign to Stop Killer Robots and Global Policy Making," *Perspectives on Politics* (Winter 2015).

Chapter Six

Terraforming: A Critical Step in the Recipe for Making Martians

THE PARADOX OF LOW MARTIAN TEMPERATURES

The Martian environment offers one of the biggest challenges to human colonization. The planet is "cold" and dry, with an unbreathable atmosphere. The average pressure of only 0.087 psi (0.006 bar) is comparable to the partial vacuum a typical high school physics teacher creates inside a bell jar when demonstrating that ice water can boil at extremely low pressures. To make things worse, the Martian atmosphere is 96 percent carbon dioxide. Even if the pressure were elevated to the same value as Earth's atmosphere (14.7 psi or 1.0 bar), the Martian atmosphere still would be unbreathable.

Explorers foolish enough to walk around on Mars's surface without the aid of a pressurized suit and an oxygen supply would last about 10 seconds before losing consciousness and about a minute before dying.[1] The lack of pressure would cause dissolved gases in their bodily tissues to fizz, forming tiny bubbles in their tissue similar to the small bubbles formed in a bottle of soda pop when it's opened. While this won't cause a person to explode or their eyes to pop out, it will make them look somewhat bloated. If the explorers were dumb enough to take deep breaths before walking outside, they might not last even 10 seconds. Without any significant outside pressure to counteract it, the air inside their lungs will expand and pop their lungs like party balloons.

Near its equator, daytime temperatures in the Martian summer can reach a balmy 70° F (20° C) but then plummet to -100° F (-73° C) at night. Unfortunately, Mars's average temperature is about -80° F (-60° C).[2] In human terms, however, the temperature does not have the same meaning on Mars as on Earth due to Mars's extremely low atmospheric pressure.

Humans are designed to lose heat continually, albeit at a fairly slow rate. If the heat loss is too slow, humans feel hot. People who are gaining heat have, at best, only a few hours to live. On the other hand, people who are losing heat faster than normal will feel cold and in the extreme case die from it rather quickly. But the heat-loss is not just a matter of temperature. On Earth, air at 80° F (26.7° C) feels warm. At the same temperature, water feels cool or for some individuals downright cold. Due to water's much higher density, a person loses heat much faster in 80° F water than in 80° F air. In other words, the greater the number of lower temperature fluid molecules in contact with a warm surface, the higher the heat loss from the warm surface to the cooler fluid.

This is the principle used in a thermos bottle to keep the soup warm. The thermos is a double-wall container, with the inner wall in contact with the soup and the outer wall in contact with the outside air (see figure 6.1).

A partial vacuum between the inner and outer walls of the thermos slows the normal forms of heat loss to almost nothing, because the density of the gas in the partial vacuum is near zero. With such a condition, very few molecules are available for transferring heat from the warm inner wall to the cool outer wall of the thermos.

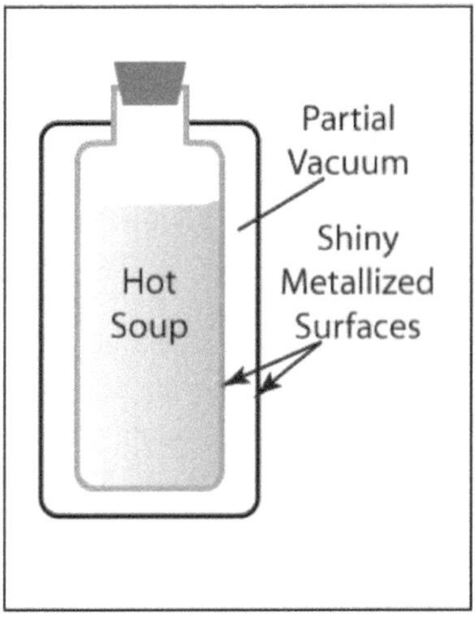

Figure 6.1. Diagram of a thermos bottle. ***Source*: T. K. Rogers.**

Of course, radiant heat transfer still can occur even in a vacuum. Here the warm inner surface would emit heat as infrared radiation. The cooler outer wall then would absorb the radiation and become slightly warmer, thereby leaking a small amount of heat to the cooler outside air around the thermos. The worst-case amount of radiation heat transfer in a thermos would be low to begin with but is reduced to almost zero by coating the thermos bottle walls with a shiny metallic coating. In Martian terms, blocking radiant heat transfer is about all one needs to do to stay warm in the planet's partial vacuum (assuming, of course, that the other needs for remaining alive are met, such as wearing a pressure suit and having a means of breathing).

Given the partial vacuum condition on Mars, an explorer walking around in a space suit will have no problem staying warm even with very cold outside temperatures. In fact, an explorer could even have trouble overheating. A stiff wind would make the heat loss from a space-suited explorer worse, but it still would be very low compared to the heat loss that would occur on Earth at the same temperature and wind velocity.

The Martian Problem of Shelter

The partial vacuum in contact with the outer surface of Martian settlements would make them almost ideal for passive solar heating. Once a structure on Mars is heated up, about the only way to lose the heat would be by radiating it back out to space as long wave-length infrared light, the same type that is mostly blocked by any form of greenhouse glazing. A thin layer of greenhouse glazing material covering a Martian settlement would allow the Sun to warm it in the daytime, yet mostly prevent the warmed surfaces from radiating their heat back to outer space at night. Of course, heat would be transferred into the ground below the settlement, but over time, the ground would warm up and the heat loss into it would decline to almost nothing.

Conceivably, a thin plastic glazing covering a Mars settlement could be inflated to form a bubble that contains a breathable atmosphere similar to the 5 psi pure oxygen atmosphere used by Apollo missions after they were launched into space. This would allow settlers to walk around outside their dwellings as long as they remained inside the plastic covering. Getting the atmosphere right in the bubble (not to mention eventually in the entire planet) would be tricky. A little too much CO_2, could result in headaches, breathlessness, visual distortion, or, in the extreme case, death. On the other hand, a little too much pure oxygen pressure, and the bubble could explode in a blazing inferno set off by an electric spark similar to what happened in the disastrous Apollo 1 fire in which three astronauts trapped inside their space capsule died in an inferno during a routine test.[3] But then the wrong level of pressure—too low or too high—by itself could cause a disaster by making the dome collapse or pop.

Many depictions of Mars settlements show groups of buildings inside large clear plastic domes. The problem is that even a dome with only a 100-foot (30.5-meter) diameter would have over 2,800 tons (25,000 kilonewtons) of upward force acting on it from its internal pressure. Keeping the dome intact and tied down is going to be a significant design and material-strength challenge.

The early Mars explorers, however, will not live in cities covered by plastic domes. They will live in prefabricated units transported from Earth, sort of like a space-age trailer park. On the other hand, considering that the price of the homes likely will be on the order of a few hundred million dollars, the early explorers will be living in the equivalent of Martian McMansions.

Given their stylish cylindrical shapes and slightly domed roofs, one can only marvel at the grandness of their curb appeal.

Unfortunately for property values, as colonization expands, there goes the neighborhood. Unlike Earth, Mars lacks magnetic poles and a thick atmosphere, both of which greatly reduce radiation exposure coming from the Sun or cosmic rays. So most Martians would need to spend a significant amount of time underground, well shielded from incoming radiation. Of course, thickening the Martian atmosphere through terraforming certainly would help, but levels of Martian surface radiation are never likely to be as low as those on Earth.

Given an initial shortage of heavy earth-moving (or should we say Mars-moving) equipment in the early days of colonization, how would the tunnels be dug? Simple: They wouldn't. After moving out of their Martian McMansions, colonists would migrate into naturally occurring caves such as lava tubes. These are formed when the outer surface of a lava flow cools and hardens. The still liquid center then drains out, leaving a horizontal tube that can be miles long. For example, the Kazumura lava tube in Hawaii has a length of over 40 miles (64 kilometers).[4]

Already a number of lava tubes have been identified from photographs of Mars's surface, some of them massive, with diameters exceeding 100 meters, possibly due to the lower gravity.[5] Include the possibility of other types of caves, and it's clear that a large population of Martians could reside in underground cities without even having to dig the caverns and connecting tunnels. The key issue would be creating and maintaining the required underground atmosphere. Cave walls might need to be sealed to prevent air loss through the porosity of the rock. Spraying sealant on walls is certainly a lot easier than blasting tunnels. Also, small lightweight pressurized habitats could be placed inside the lava tunnels for radiation protection without pressurizing the tunnels themselves.

Undoubtedly, over a period of decades, Martians would add significantly to their underground habitat by digging their own tunnels, even if they had to do so with pickaxes and shovels. History gives many examples of massive tunnel systems dug with relatively simple equipment, such as the tunnels dug by Communist forces during the Vietnam War.[6] These provided living space for soldiers, storage space for supplies, and a means of transporting troops and equipment in a stealthy manner secure from bombardment. Martian cities are going to be a labyrinth of tunnels and caverns.

Those underground cities could be warmed by passive solar energy collected in greenhouses placed above the ground structures. Martian domiciles are not going to be architectural wonders that instantly communicate the wealth and status of the occupants to passing motorists (not to mention that the "motorists" will, for the most part, be traveling in tunnels). But building large underground settlements on Mars is not going to be all that difficult compared to building massive skyscrapers and subway systems on Earth. Ironically, one of the most technologically advanced societies produced by humanity will be made up of cave dwellers.

The Problem of Providing Food

Martian greenhouses could be small-sized plastic domes or long, inflated translucent plastic tubes similar to the plastic-covered high tunnels commonly used as greenhouses on Earth (see figures 6.2 and 6.3).

The Martian greenhouse tubes would be large enough in diameter to walk upright in but unlike the high tunnels on Earth, which are built by covering semicircular hoops with translucent plastic film, the Martian version would be tubes of plastic film. The bottom of the tubes would contain the growing medium—either some form of hydroponic matter or actual soil. The enclosed tube shape would solve the problem of the upward force that would exist with a dome. The downward pressure on a tube's bottom would counteract the upward pressure on its top side while the growing media on the bottom of the tube would anchor it in place. The tube also would prevent the loss of precious water or gas pressure to the Martian soil below the tube or to the Martian atmosphere above.

Of course, the use of such greenhouses assumes that the plants grown in them would not be seriously harmed by the higher levels of radiation bombarding the surface compared to Earth. A thin plastic covering would offer little shielding. Plants can develop cancerous tumors, so some amount of care

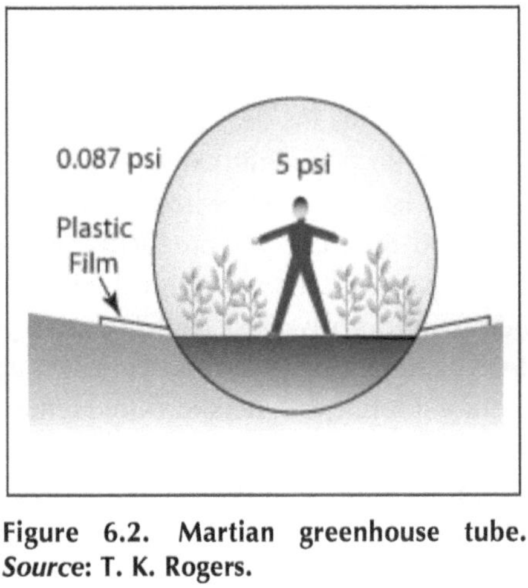

Figure 6.2. Martian greenhouse tube. *Source*: T. K. Rogers.

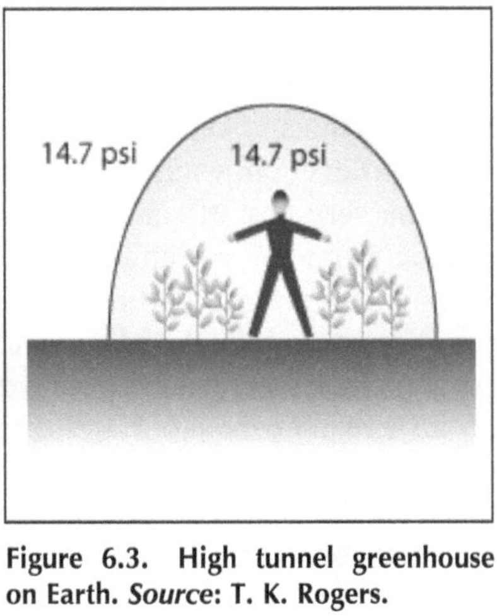

Figure 6.3. High tunnel greenhouse on Earth. *Source*: T. K. Rogers.

in plant selection and/or genetic engineering might be required. Luckily, most food crops are grown as annuals and only would be exposed to the radiation for a few months. By comparison, humans foolish enough to live on the surface of Mars without radiation shielding would be exposed for decades. Radiation-sensitive crops could be grown in underground tunnels using artificial light or light reflected into the tunnels using mirrors.

The infrared-blocking property of the plastic combined with the outside partial vacuum condition would prevent most of the nighttime heat loss. If a single layer of plastic failed to retain enough heat, a second layer could be added to the outside with a gas space between them. The gas space would be near the outside pressure in order to insulate against heat loss but a little higher than the outside pressure in order to keep the layers apart. Possibly a type of transparent bubble wrap could also be used as insulation.

Heat loss into the exposed Martian soil at the edges of the tunnels could be solved by covering a few feet of the soil on each side of the tunnel with the same type of plastic film used in the greenhouse to warm it and insulate it from nighttime radiant heat loss. The film would be set above the surface so that it would trap a thin layer of CO_2 between it and the soil. This layer would be only slightly higher in pressure than the outside atmosphere, just enough to hold the plastic film above the soil. Possibly, the film would be metallized to create a shiny surface for reflecting additional light into the greenhouse tube.

The lower incident light levels on Mars could be alleviated using reflective films as mentioned above. A tube-shaped greenhouse placed on the surface also would receive a significant amount of diffuse light reflected off Mars's surface. This reflected light would have a red tint, but then red light is generally useful for plant photosynthesis.

Even on Earth, the high tunnel–type greenhouses can have problems overheating during the day and have to be ventilated. With the low daytime heat losses on Mars, greenhouse tunnels located near the equator would almost certainly overheat. But the solution is simple: Vent the heat into nearby tunnels and habitats to keep them warm for the humans inside.

Food production would be a limiting factor on population size. On Earth, under the very best of conditions, at least an acre of land is needed to feed a person (with a very limited diet). The land area per person for the typical American diet would be much higher, thanks partly to a high intake of animal-derived products. Mars would be able to grow food using relatively simple passive solar-heated plastic domes or tube-type greenhouses. Even at only an acre per person, it is going to require a lot of thin plastic film. Hence, the ability to expand the population may depend on the ability to manufacture thin plastic film on Mars itself. Importing film from Earth for anything other than a few explorers or early colonists would be quite expensive.

Using Greenhouse Gases for Warming Mars

Increasing the atmosphere's pressure on Mars would help counteract the upward force on plastic domes, making it possible eventually to cover entire settlements or build giant greenhouses. But it also would increase the heat loss from such structures to the outside atmosphere, that is, if the outside temperature remained the same. The good news: Increasing the planet's temperature is the means for increasing the atmospheric pressure, not to mention making liquid water more readily available and increasing the possibility of growing plants on the surface without the need for greenhouses.

If the planet's temperature were increased, a significant amount of CO_2 would outgas from being frozen in the soil, and the polar caps, consisting of frozen CO_2 and water, would largely melt. The thickened CO_2 atmosphere would create a greenhouse effect on the planet leading to a self-sustaining warming trend. It is conceivable that this melting of frozen CO_2 could raise the atmospheric pressure to around 5 psi, the same pressure used inside possible dome cities, thereby simplifying their construction.

Mars's self-sustaining warming process could be initiated by artificially injecting powerful greenhouse gases into the planet's thin atmosphere. As greenhouse gases go, CO_2 is by far the most common one in Earth's atmosphere but by no means the most effective one. CO_2 is given a greenhouse gas potential (GPW) of 1.0 by the Environmental Protection Agency. By comparison, methane has a GPW of 28 to 36 and lasts about a decade on average. Nitrous oxide (N_2O) has a GWP 265–298 and remains in Earth's atmosphere for more than 100 years, on average. Chlorofluorocarbons (CFCs), hydrofluorocarbons (HFCs), hydrochlorofluorocarbons (HCFCs), perfluorocarbons (PFCs), and sulfur hexafluoride (SF_6) have GPWs in the thousands or tens of thousands.[7] Obviously, if one wants to deliberately create global warming on Mars, then clearly, the various fluorinated compounds are the way to go. Injecting even a few hundred parts per million of fluorinated gasses into the atmosphere would likely be enough to produce a substantial amount of atmospheric warming but would take decades or even centuries.

Fortunately, the materials for making fluorinated compounds are all available on Mars. The manufacturing equipment and energy sources needed to power it are not. Even using advanced manufacturing techniques and robots, warming Mars by injecting greenhouse gases into its atmosphere is probably not going to be realistic in the early days of exploration and colonization. Then again, a significant amount of colonization could be done even with the current thin atmosphere. Having a thick atmosphere with a warm climate is not as critical to colonization as might at first be assumed, as long as water needs can be met.

Obviously, agriculture, let alone keeping people hydrated, is not going to be possible without a source of liquid water. Perhaps this source could come from wells tapping into possible underground aquifers. Schemes for collecting water out of the Martian atmosphere have been proposed.[8] It turns out that the thin Martian atmosphere approaches 100 percent relative humidity at nighttime. By passing atmospheric gases through a special metal-organic framework (MOF) the water vapor can be absorbed. On the following day the vapor can be removed from the MOF as liquid water using solar energy. Melting possible ice deposits found in caves also may be an option. Making liquid water widely available on the surface, however, ultimately will require a thicker, warmer atmosphere.

The right type of bacteria could be used for creating greenhouse gases like methane or ammonia without the need for chemical manufacturing equipment and power sources, but bacteria need a food source. This could come from human waste or agricultural waste, such as the otherwise inedible leaves and stalks of crops. Unlike on Earth where such things are considered problems, agricultural and human waste are going to be highly valuable materials on Mars, probably too valuable to be used just for feeding greenhouse gas–producing bacteria.

Human and agricultural wastes would be invaluable for creating fertility in Mars' otherwise sterile soil, in order for the soil to act as a medium for producing protein. For the most part, Mars will have no protein-providing livestock because there will be no pastures or vast crops of corn to feed them. Instead, a significant amount of protein could be provided by growing various types of mushrooms on the human and agricultural waste products. And while mushrooms in the United States often are considered nothing more than a pizza topping, in other cultures they are considered a common food source and gourmet delight. In the process of growing, the mushrooms also will help convert the waste products into compost to be used for creating soil fertility. What's more, mushrooms could be grown underground in low light conditions.

While these and other modifications are enough to make Mars livable if residents remain in their habitats, additional modifications will be needed if Mars is to become at least somewhat Earth-like.

Crashing Comets

Warming up Mars also could involve colliding water-bearing comets and/or asteroids with the planet's surface. Demonstrating the ability to control the movement of such items has profound implications for space mining and the protection of Earth from devastating impacts. On Mars, the collisions would both warm the surface and provide additional greenhouse gas such as ammonia (often found in asteroids) or water. Ceres, the largest asteroid in the asteroid belt, alone is thought to contain more water (typically as ice) than is found on Earth.[9] While Mars is thought to have a significant supply of water that could be released by warming up the planet, the supply may not be enough to support anything other than a desert environment. Bringing in additional water could be important.

An asteroid with a diameter of 130 meters slamming into Mars at about 20,000 miles per hour would release about as much energy as a 250-kiloton thermonuclear bomb (assuming a density of 2000 kg/m).[3] This is about the same energy as a typical city-destroying warhead carried by an intercontinental missile. An asteroid 32 meters in diameter with the same density and speed at impact would release about the same energy as the bomb that destroyed Hiroshima in World War II. Cause it to crash into one of Mars's polar regions, and it undoubtedly would vaporize a significant amount of otherwise frozen CO_2 or water, thereby releasing it into Mars's atmosphere.

While moving a large chunk of mass through the vastness of outer space so that it collides with a planet sounds impossible, it is actually feasible. Mars, after all, is located near the asteroid belt. A relatively small-sized nuclear-powered drone could attach itself to a nearby piece of space junk and use ion propulsion to slowly alter the object's orbit enough to put it on a collision path with Mars. Keep in mind that a 130 meter–diameter asteroid would have a mass of only about 27 million kilograms as compared to a *Nimitz*-class aircraft carrier's mass of 93 million kilograms. While the aircraft carrier's nuclear power system can only propel it at a cruising speed of 30 knots (56 km/h)—pathetic by space-travel standards—it also has an enormous resistance force acting against it since it is moving through water. In the near perfect vacuum conditions of outer space, there is no resistance force. Moving a 130-meter piece of space junk would be far easier to do than moving an aircraft carrier through Earth's oceans.

Crashing large pieces of space junk into Mars would pose some degree of risk to its human inhabitants, but even with a human population in the millions, their cities would be tiny targets compared to the entire surface of Mars. Martians probably would start by crashing relatively small-sized objects that posed no more risk than having an out-of-control space vehicle crash when attempting to land near a settlement. As the asteroid crashing technology improved, the size of the objects could be increased.

The whole process could be automated with robotic space drones so that an almost continual stream of space junk would be crashing into remote areas of Mars. While the water or greenhouse gases added to Mars from a single impact would be minor, decades' worth of impacts would be substantial. And it could be done without risking major problems of dust creation or seismic activity. If the pieces of space junk were the right size and composition, they would explode in the atmosphere rather than creating impact craters.

Of course, Martians simply could use hydrogen bombs to defrost the planet instead of going to all the bother of crashing asteroids that release a similar amount of energy. The dangers of radiation would be an issue, but if the bombs were detonated high above the surface in remote areas, the radiation problem could be minimized. Nevertheless, the bombs would add no additional water or greenhouse gases to the planet. Also, it is unlikely that Earth's nuclear powers would happily donate a supply of bombs.

Enormous space-based mirrors also could be used to reflect light onto the polar regions of Mars. Since light exerts pressure on the objects it shines on, albeit a minuscule pressure, the light hitting the mirror conceivably could be used to counterbalance the gravity pulling the mirror toward Mars's surface. Hence, mirrors conceivably could be held in fixed positions to reflect on Mars's polar caps.

Oxygenating the Atmosphere

During the terraforming process, the Martian atmosphere would evolve from nearly pure CO_2 at partial vacuum conditions to CO_2 at about 5 psi (1/3 bar). While the planet could support a significant human population even under the partial vacuum condition, once the higher atmospheric pressure was reached, actually making the atmosphere breathable would be the next logical step. To do so, the vast majority of the carbon in the CO_2 would need to be removed from the atmosphere and the oxygen released.

In the early days of colonization, much of the oxygen in human habitats would be produced artificially from CO_2 using various chemical processes. As greenhouses were expanded, plants would play an increasingly important role. Beyond the greenhouses and human habitats, conditions would be extremely unfavorable for most forms of plant life. Fortunately, studies on Earth have shown that at least some forms of lichens and cyanobacteria not only can survive but actually grow in Mars-like conditions.[10] Furthermore, they actually can produce oxygen from CO_2 and slowly convert rock, gravel, and sand into soil for growing plants.

It would be much easier to make air breathable by filtering out an undesirably high level of CO_2 if it contains mostly oxygen rather than chemically making oxygen out of CO_2. Portable equipment to do the filtering would weigh far less than carrying around an oxygen supply in pressurized tanks. So using lichens, cyanobacteria, and eventually plants to enrich the oxygen content of the atmosphere would be a worthy goal, albeit one that could take centuries to complete.

Despite significant challenges, Mars is the only place besides Earth in the solar system with enough gravity and other resources required conceivably to create a human-friendly atmosphere, which makes it particularly attractive for colonization. Even if Martians had to carry breathing apparatuses to walk around on their planet's surface, not having to cope with extreme vacuum conditions and having even a partial oxygen atmosphere would be definite benefits.

Mars may in fact be "cold as Hell," but with a little human ingenuity and hard work it could be the "kind of place to raise your kids"—provided you can get a reliable source of energy.

NOTES

1. Geoffrey A. Landis, "Human Exposure to Vacuum," accessed June 5, 2019, http://www.geoffreylandis.com/vacuum.html.
2. Tim Sharp, "What Is the Temperature of Mars?," accessed June 5, 2019, https://www.space.com/16907-what-is-the-temperature-of-mars.html.
3. Matteo Emmanuelli, "The Apollo 1 Fire," accessed June 5, 2019, http://www.spacesafetymagazine.com/space-disasters/apollo-1-fire/.
4. Joshua Foer, "Inside the Deep Caves Carved by Lava," accessed June 5, 2019, https://www.nationalgeographic.com/magazine/2017/06/lava-tubes-hawaii-caves/.
5. Anita Heward, "Lava Tubes as Hidden Sites for Future Human Habitats on the Moon and Mars," accessed June 5, 2019, https://phys.org/news/2017-09-lava-tubes-hidden-sites-future.html.
6. Lincoln Riddle, "The Cu Chi Tunnels: A Dangerous Underground Warzone in the Vietnam War," accessed June 5, 2019, https://www.warhistoryonline.com/vietnam-war/cu-chi-tunnels-dangerous-underground-warzone-b.html.
7. Environmental Protection Agency, "Understanding Global Warming Potentials," accessed June 5, 2019, https://www.epa.gov/ghgemissions/understanding-global-warming-potentials.
8. Peter Rejcek, "A Mars Survival Guide: Finding Food, Water, and Shelter on the Red Planet," accessed June 5, 2019, https://singularityhub.com/2017/05/28/a-mars-survival-guide-how-to-find-food-water-and-shelter-on-the-red-planet/#sm.00001tfrpjmixfcwzqz741lzt31f.
9. Michael Küppers, Laurence O'Rourke, Dominique Bockelée-Morvan, Vladimir Zakharov, Seungwon Lee, Paul von Allmen, Benoît Carry, David Teyssier, Anthony Marston, Thomas Müller, Jacques Crovisier, M. Antonietta Barucci, and Raphael Moreno, "Localized Sources of Water Vapour on the Dwarf Planet Ceres," *Nature*, vol. 505 (January 23, 2014): 525–27.

10. Mike Malaska, "Earth's Toughest Life Could Survive on Mars," Planetary Society, accessed June 5, 2019, http://www.planetary.org/blogs/guest-blogs/20120515-earth-life-survive-mars.html.

Chapter Seven

Energy Production: The Nuclear and Solar Options

PROBLEMS WITH SOLAR PANELS

Of the four sources of wealth, energy supply is the one most critical to extended habitation of Mars. Energy is required for climate control, oxygen production, food cultivation, lighting, and all other electric power needs. Given the length of supply lines, it is impossible to imagine a colony surviving without a locally available supply of energy.

Thanks to images of Mars rovers driving about mobilized by photoelectric solar panels, those panels immediately pop into mind as the solution for powering Mars, but keep in mind that the Mars rovers only moved a few hours a day and then at slow speeds. Their solar cells could output about 140 watts for about four hours a day, not very impressive compared to the 100,000 watts available to power a typical car on Earth. Then there are the problems of dust storms blocking sunlight, the accumulation of dust on the collectors, and the considerable reduction of incident sunlight the farther one travels away from the equator.

The latest Mars rover employed nuclear rather than solar power to overcome the above problems. The nuclear device used plutonium 238, an isotope that decays fast enough to make it glow red. A compact thermoelectric generator then converts the heat into electricity without the use of moving parts. The nuclear unit in the Curiosity rover only outputs about 110 watts, but it does so in the dark of night and independent of dust storms and accumulated dust. Hence, Curiosity's nuclear power device can provide about 2.5 kilowatt hours of energy per day compared to the earlier solar-powered rover's mere 0.6 kilowatt hours per day.

Curiosity's nuclear power source sounds awesome until one considers that 110 watts is not a lot of power, not to mention that the cost of plutonium 238 is about $8 million a kilogram. In the past, it was man-made as a by-product of nuclear weapons manufacturing. But today, pretty much no one still makes it. Obviously, this type of electrical generation is not going to power a Mars settlement.

Due to its increased distance from the Sun, Mars receives about 44 percent as much solar energy per unit of area as Earth as measured before the sunlight passes through the respective atmospheres of the two planets. More sunlight is absorbed in Earth's thicker atmosphere than in Mars's thin one. So on the surface at the equator during the brightest time of day, Mars receives about 60 percent of the amount of sunlight that Earth receives under the same conditions. The solar light intensity at the equator on Mars is similar to the solar light intensity in the state of Virginia. Hence, solar panels would work on Mars, as long as one were reasonably close to the Martian equator.

Unfortunately, a lot of hardware is required for powering a settlement with solar panels, especially considering the massive battery storage needed for providing power at night. Sending this much hardware from Earth to Mars would be horrendously expensive. Manufacturing the equipment on Mars would be a solution, but then Mars would need to have a high-capacity source of electrical power (with a lot of other equipment) in order to manufacture even a single solar panel.

Then there is the reliability of solar panels, considering that Mars can have massive sunlight-blocking dust storms lasting as long as a year. A serious reduction of power on Mars would not just be annoying; it would be life-threatening. First and foremost, there is the problem of breathing. After decades of terraforming, Mars might have an oxygen-rich atmosphere dense enough to be breathable, but in the early days it would not. Oxygen for Martian habitats likely would be produced using electricity to break apart CO_2 molecules into O_2 and CO (with a process for removing the carbon monoxide). Hence, less electric power eventually would mean less breathing.

Should it get a little chilly in the habitat at nighttime, it would not be possible to throw another log into the woodstove—there would be no logs. There would be no natural gas for cooking or heating water. In fact, there would be no kerosene for lamps or even paraffin for candles. Without electricity, human life on Mars (if it could continue even to exist) would be worse off than on preindustrial Earth.

There is also the interaction of electrical power with food production. In the early days of colonization, food production would depend on some form of aquaculture, hydroponics, or possibly spirulina growing system. Typically, these systems need electric-powered pumps to move water around or compressors to bubble air into the water used for growing. They also might require artificial lighting to maximize growth, especially during events like major dust storms that reduce available sunlight in greenhouses—just the times when electricity production is lowest. Plants grown in soil would not require the pumps but would still need some amount of artificial light. Then there is the problem of temperature control in greenhouses. Even if they were designed well enough to stay warm at night, during the daytime, without ventilation, they could overheat. A lengthy loss or reduction of electrical power could result in crop failure.

There is a way to alleviate the problems of generating Mars's electrical power from solar panels: Place the panels in orbit and transmit the energy to Mars's surface. The panels could be placed in geostationary orbits, meaning they would continually remain over a particular location on the Martian equator. The solar power system would complete its orbit in exactly the same time as Mars completed a revolution. Electrical power would be transmitted to the planet via microwaves. Power would be interrupted only when the solar panels were in the shadow of the planet. This could be overcome by having multiple satellites equipped with solar panels. Some amount of battery storage still would be needed to compensate for variations in electrical demand, but the battery requirements would be far less than when using ground-based solar panels.

Orbiting solar panels have been proposed for Earth but so far have not been pursued due to cost, including the considerable cost of launching the equipment.[1] Equipment for the original solar panel systems orbiting Mars would need to be manufactured on Earth and not only be launched into space from Earth's surface, but then require additional fuel to make the rest of the trip to Mars. Such a power system would be expensive, but considering it cost $2.5 billion for the Curiosity robot mission that sent an underpowered robot driving a few dozen miles around Mars, building an expensive power system that enables colonization of a new world seems reasonable.

Why Mars Is a Tough Energy Start-Up

On Earth, some photoelectric solar panel shortcomings can be alleviated with the addition of wind turbines, another form of solar energy collector in which large parts of a planet's surface collect the energy that creates the wind. This reduces the need for batteries since the wind can blow at night. Unfortunately, the density of the atmosphere on Mars is so low that a hurricane-velocity wind creates very little force on a structure. The scene in the movie *The Martian* in which the rocket is about to be tipped over by wind is pure fiction.

While a wind speed of 23 mph is considered sufficient for generating electricity with wind turbines on Earth, initial research indicates that a speed of 67 mph (nearly three times higher) would be required to get a similar performance from wind turbines designed for Mars. The good news: Such wind speeds do occur during Mars' massive dust storms—occasions when solar cells will not properly function. So, wind turbines may be useful as a backup system for solar panels.[2] At normal wind speeds, Martian wind turbines will have a very limited power output, at least until the atmosphere is thickened by terraforming.

Wind tunnel experiments and theoretical work have indicated that some power can be generated even at the normal low-velocity winds on Mars—enough power to run at least some of the equipment on a future Mars probe.[3] However, as mentioned earlier, operating a Mars probe requires a very small amount of power. Questions remain about whether these lower-wind-velocity turbines could be scaled up and produce enough energy to be useful for powering a Martian city.

In the early stages of colonization, Martians may be able to exist by using solar panels, possibly with wind turbine backup during dust storms. However, this would be an austere existence. Self-sufficiency would require manufacturing and heavy industries such as steelmaking. These will likely require additional power sources.

Aerospace engineer Robert Zubrin, the driving force behind the Mars Direct proposal, has suggested that Mars could be powered with geothermal energy, but this would require exploration and drilling. Again, something has to power the process of developing geothermal (if it can be developed) before it comes online.[4] This brings up the concepts of EROI (energy return on investment) and uncertainty. Expending the energy to identify and develop new energy sources only makes sense if the certainty of development and the EROI are favorable. At this point, geothermal has a high level of uncertainty.

Starting up a Martian energy supply system will be a challenge. By contrast, if Earth were an uninhabited planet, finding temporary energy sources for a colony in the startup phase would be straightforward for high-tech colonizers. They could start with a campfire. Earth has lots of oxygen in its atmosphere and lots of combustible materials. Without oxygen, fossil fuels and firewood are useless. But the oxygen in Earth's atmosphere is also conveniently diluted with nitrogen, which keeps fuels from exploding violently when combusted. Early explorers on Mars couldn't barbecue if they wanted to. Fossil fuels would be useless, even if they bubbled out of the ground. On Mars, this inability to use combustion would greatly complicate otherwise straightforward tasks.

There is a reason why heavy construction equipment on Earth typically runs on diesel fuel. It has a very high energy density (especially compared with batteries) and a fast refueling rate. In an environment with no oxygen, such a vehicle also would need to carry a liquid oxygen tank. Combined with the mass of the diesel fuel, the total would be five times more mass than the diesel fuel alone. This essentially would be running the vehicle on rocket fuel. If the vehicle carried an air tank with a nitrogen content similar to Earth's atmosphere, the total fuel plus air would have 18 times more mass than the diesel fuel alone, and that's assuming that no excess oxygen was needed for complete combustion—yeah, right.

Without an oxygen-rich atmosphere, heavy construction equipment for Mars would pose a major design challenge. Without such equipment, the development of Mars would be limited. Due to the lack of a thick atmosphere and a magnetic field, the surface of Mars gets significantly more ionizing radiation—the kind that is harmful to life—than on Earth. Martian cities would need to be at least partially underground to compensate for these differences. This suggests the need for heavy dirt-moving equipment, similar to the diesel-powered monsters used on Earth. While some level of photoelectric solar panels could be installed with hand tools, creating the power plants using geothermal or nuclear power required for powering construction and eventually manufacturing is going to require heavy construction equipment: cranes, bulldozers, excavators, and so forth.

The Nuclear Option

Ultimately, Mars colonization will require nuclear power, which conjures images of gargantuan reactor containment buildings and cooling towers, not exactly an ideal payload for launching into space. To make things worse, the light water uranium-based reactor types currently used for electrical generation on Earth need, surprise, surprise, water—a quantity in short supply on Mars, at least in liquid form. The water is required for cooling the reactor's core. Stop the flow of cooling water on the core, and it melts into a radioactive calamity. To prevent this from happening, Earth's nuclear power plants typically have backup generators powered by fossil fuels—also not available on Mars—to ensure that cooling water pumps keep running even after a natural disaster like a tsunami (hopefully unlike the tsunami that wiped out the backup systems and caused the Fukushima Daiichi reactor meltdowns).

There is more than one form of light water reactor, but for all of them the term "cooling" water, as applied to the reactor's core, is a paradox. For example, the most common light water reactor type, the pressurized water reactor (PWR), has a core cooling water temperature of around 617° F (325° C), not exactly what normally would be considered cool. Still, the water flow does keep the core from overheating. To keep the water from flashing into steam requires a pressure of around 2250 psi, or about 150 times higher than Earth's atmospheric pressure.[5] Hence, the reactor core has to be housed in a huge steel pressure vessel. A lower core cooling water temperature could use a lower pressure and lighter weight pressure vessel, but as will be shown in the discussion below, the lower temperature would yield unacceptably poor efficiencies when the reactor's heat is converted into electrical energy. Needless to say, shipping huge steel pressure vessels from Earth to Mars is hardly ideal.

The PWR has become Earth's basic design partly because it was well suited for use in submarines. As a side benefit, it produced plutonium that could be used in nuclear bombs. What could be better? Produce materials to make warheads for blowing up enemy cities, and provide a way to keep the warheads hidden by stealthily moving them around under the ocean ready for launching, all with just one reactor type.

Thanks to the hard-driving tutelage of Admiral Hyman G. Rickover, the U.S. Navy used PRW technology created during the Manhattan Project to produce some exceptionally well-engineered reactors for its submarines and eventually its aircraft carriers. In the process, the navy also produced a well-trained and well-vetted workforce to design, maintain, and operate the reactors. So when the electric utility industry became interested in nuclear power, the decision to use PRW technology rather than taking a multibillion-dollar gamble on something less well developed but possibly better was pretty much a no-brainer.

A submarine-sized reactor might be usable on Mars, but a more compact unit designed specifically for Mars would be even better. Fortunately, other more compact designs already have been demonstrated in the earlier days of nuclear power development, although they were not generally pursued into commercial development.

Not to be left out, the U.S. Air Force had enlisted the services of the Oak Ridge National Laboratory to create a reactor for powering a bomber. Obviously, this reactor would have a compact, transportable design. Although in tests it did not run the aircraft's six propellers and four jet engines, the reactor was actually powered up while being flown inside a massive B-36 bomber. The crew section of the bomber was outfitted with 11 tons of shielding, and 47 flights were made with the reactor on board in order to evaluate whether the bomber's crew could be safely shielded from radiation.[6] To save weight, nothing else on the aircraft was shielded, so everything on the ground was, to some degree, depending on the aircraft's altitude, irradiated as the bomber flew overhead.

Intercontinental ballistic missiles—a faster unmanned system with a nearly 100 percent chance of payload delivery—eventually made the nuclear-powered bombers obsolete, not to mention that there were concerns about radioactive contamination if a nuclear-powered plane crashed and, yes, even some concern about irradiating people in the bomber's flight path. The project was canceled in 1961.

Unlike the PWRs, the nuclear aircraft engine used no water or solid fuel and achieved a maximum outlet temperature of slightly over 1500° F (820° C).[7] A molten uranium-containing salt acted as the fuel and was circulated as the equivalent of core coolant. Since the salt had a very high melting temperature, there was no need to maintain it under high pressure to keep the salt from flashing into vapor, thus eliminating the need for a heavy pressure vessel to contain the reactor core.

Thorium to the Rescue

Like many government-funded programs, the nuclear-powered aircraft program didn't actually die just because it was canceled. It simply refocused itself on creating a safer meltdown-proof reactor for electrical power generation. The work included building and running a Molten-Salt Reactor Experiment (MSRE) at Oak Ridge beginning in1965. This laid the groundwork for developing the LFTR (pronounced lifter), for Liquid Fluoride Thorium salt Reactor. It too would use a molten salt for fuel and core cooling, except this time the molten salt would be thorium fluoride. With a melting point of 1200° F (649° C), there would be no need to keep the molten salt under high pressure nor any need for an enormous pressure vessel.

The MSRE was shut down in 1969 by government bureaucrats to help provide funding for developing the uranium-based liquid metal cooled fast breeder reactor (LMFBR). The liquid metal used as a coolant in this case was liquid sodium, a coolant that would burst into flames if it leaked out and contacted air or explode if it contacted water.[8]

At the time, the LMFBR was considered the energy savior of humanity by no less than that famous "physicist" President Richard Nixon, who confessed during a speech at the Atomic Energy Commission's Hanford Works:

> all of this business about breeder reactors and nuclear energy is over my head. That was one of my poorest subjects—science. I got through it, but I had to work too hard. I gave it up when I was a sophomore.[9]

As was often Nixon's habit, he was kind enough also to leave a recording of a private telephone conversation with a political crony discussing how his home state of California deserved to be blessed with the benefits of the program's government spending.[10] But it wasn't just political maneuvering; Nixon seemed to actually believe the LMFBR was the next big thing.

While most of the world's supply of uranium was in the form of U^{238}, U^{235} was the isotope required for running nuclear reactors or making atomic bombs. Since U^{235} is typically only 0.711 percent of the mass of naturally occurring uranium, its content has to be increased to between 3 percent and 5 percent for nuclear reactors.[11] For the atomic bomb dropped on Hiroshima, the U^{235} content was increased to 80 percent.[12] For more modern atomic bomb devices, the U^{235} concentration is typically in excess of 90 percent. On paper, the LMFBR was capable of converting much of the heretofore unusable U^{238} into nuclear fuel, which would have provided the United States with an almost unlimited supply of energy. What could possibly go wrong? Well, unfortunately, a lot.

The program culminated in the multibillion-dollar construction and startup of the Fermi 1 LMFBR reactor located near Detroit, which soon after, experienced a partial meltdown in 1966. It took four years to repair the damage, only to decommission the plant in 1972. While the claim that the partial meltdown came dangerously close to making southern Michigan uninhabitable mostly has been dismissed by official sources, it was described in horrifying detail by a mostly forgotten book, *We Almost Lost Detroit* by John G. Fuller.[13] It has also been celebrated in song and dance by numerous video versions of the 1977 hit song "We Almost Lost Detroit" by Gil Scott-Heron.[14] Whether the partial meltdown was or was not a near human disaster, it certainly was an economic one. The LMFBR program has, as a result, passed largely into obscurity much like the rather terrifying message in the song.

Since the LFTR's fuel is already molten, having a meltdown is a non-problem. If the reactor temperature goes out of control in an LFTR, the only additional thing to melt is a freeze plug–like feature. When the freeze plug melts, the fuel drains into a specially prepared holding tank and the nuclear reaction stops.

While improved safety is a big advantage, the benefits of the LFTR go beyond safety. Given the higher operating temperature (typically 1200+° F), the efficiency of converting reactor heat into electrical energy would be significantly better for the LFTR than the PWR-type system. A nuclear reactor by itself only produces heat. To get electrical energy, the heat has to be transferred to a working fluid, such as steam, and the fluid run through a heat engine (generally a form of turbine) that turns a generator to produce electricity. The second law of thermodynamics tells us that even if there were no frictional losses, far less than 100 percent of the heat can be converted into the mechanical energy for turning the generator (see figure 7.1)

In the real world, which includes friction and unwanted heat losses, only about 38 percent of the heat generated by a PWR is successfully converted to mechanical energy by the turbine. In addition, the generator has losses due to its electrical resistance, and the reactor itself consumes a certain amount of power to run its pumps and an assortment of other support equipment.

The higher outlet temperature of the LFTR would enable its heat engine to convert about 45 percent of the heat generated by its nuclear reactor into mechanical energy for turning a generator using a closed-cycle turbine. What's more, the cooling requirements of an LFTR are about half of the requirement for a conventional plant.[15] This is yet another factor that makes the LFTR attractive for powering Mars, since Mars would not have the massive amounts of cooling water normally used with a PWR.

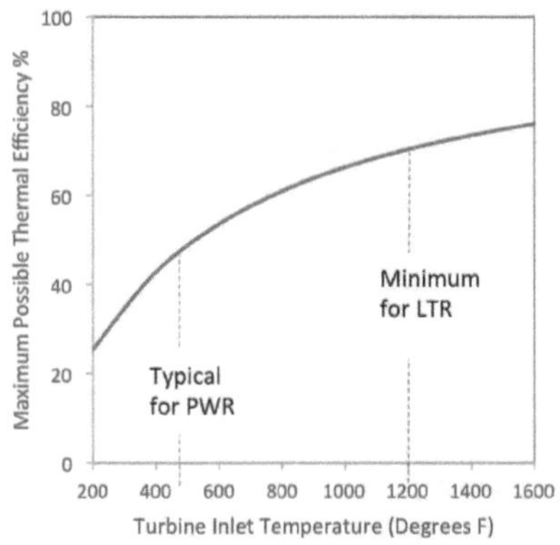

Figure 7.1. Maximum possible efficiency (Carnot efficiency) of a turbine versus its inlet temperature (assumes an outlet temperature of 32°F). ***Source*****: T. K. Rogers.**

As yet another benefit, the amount of nuclear energy available for producing heat would be much higher in a thorium-based LFTR than a conventional uranium-based PWR. PWRs typically convert about 0.5 percent of the energy available in the fuel—a mix of U^{238} and U^{235}—into heat, while the LFTR can convert nearly100 percent with less than 1 percent of the long-lived radioactive waste.[16] Thorium is also about four times more common on Earth than uranium. While it's hard to say if this concentration will be the same on Mars, initial surveys indicate there is a significant supply of thorium there. Although as yet to be proven, thorium also may be a by-product of asteroid mining. So once the system was developed on Mars, there would be no reason to import nuclear fuel from Earth.

The Downside of Thorium

Thorium has to be irradiated with neutrons in order to set off a chain of events allowing it to become fissionable and produce power. On Mars, this likely would need to be done with enriched uranium supplied from Earth. Keep in mind that just enriching enough uranium to make a nuclear bomb in World War II consumed a massive amount of resources in the Manhattan Project. However, once an LFTR was in operation, it would not need to be repeatedly jump-started with uranium.

The main barrier to the use of LFTRs on Mars stems from the fact that even though a considerable amount of development was done on the LFTR at

Oak Ridge Laboratories, the LFTR was never fully demonstrated. Technical problems such as corrosion still need to be solved.

The cost of developing the current PWR nuclear technology was substantial. Historically, such spending has been driven by fear: first, fear in World War II that the Nazis could develop an atomic bomb before America, which led to the initial development of nuclear energy technology. Then fear in the Cold War that the Russians could annihilate us before we could annihilate them, which led to the development of nuclear power in submarines. Generating electricity with nuclear power for civilian use was a secondary thought designed to help ensure public support. Nonetheless, once the navy developed nuclear reactors for submarines, taking advantage of the same technology for electrical generation was a logical step. While the cost of completing development of the LFTR would be less than developing the PWR technology, it still could be significant.

Another serious meltdown incident in an existing nuclear plant could provide plenty of fear to motivate spending, but instead of motivating a call for a safer nuclear-power plant replacement, it could motivate a call to eliminate all forms of nuclear power plants. Global warming might provide motivation for LFTR development, but global warming is, at best, a slow-motion disaster. These don't provoke knee-jerk reactions for massive spending as do moments of pure terror. Furthermore, the idea that global warming can be alleviated by solar and wind power already has been planted in the public consciousness.

Still, there may be a golden opportunity for LFTR development hidden in the rise of technologies addressing global warming. Photoelectric solar cells, high-strength magnets in wind turbines, and electric car batteries all depend on rare earth metals. For example, there are 20 pounds of rare earth metals in the batteries of a single Toyota Prius.[17] The exploding market for rare earth metals is one of the motivations for asteroid mining. On Earth, China has virtually cornered the market for rare earth metals. While there are rare earth mines in the United States, these are at a major disadvantage to China's mines. Thorium is a significant by-product of rare earth mining. Since thorium is radioactive, albeit mildly so, by U.S. law it must be reburied as a low-level radioactive waste. Imagine the impact on rare earth metal supplies if a major market existed for thorium—one that would provide a virtually inexhaustible supply of greenhouse gas–free energy with very little radioactive waste to be disposed.

A single U.S. mine's yearly thorium output could power the entire world for a year. Thorium is so energy dense that a lifetime supply could fit in a person's hand and would be smaller than a golf ball. (Imagine carrying around a lifetime supply of gasoline, not to mention the lifetime supply of air required for combustion).[18] By contrast, if all of a person's electrical power used in a lifetime were produced by a conventional nuclear reactor, just the radioactive component of the nuclear waste would fit in a soda pop can—that is, if the spent fuel were reprocessed to separate out the radioactive component. Without reprocessing, the waste would fill 33 soda pop cans.[19] Unfortunately, reprocessing efforts have been shut down in the United States.

Even if the United States did not choose to fund the development of LFTR systems to use its supply of thorium, both India and China are moving in that direction. If the American industries dependent on rare earth metals become desperate enough, they conceivably could start lobbying the federal government to fund LFTR development in order to make use of their thorium by-product.

The United States alone has poured billions of dollars into the quest for a hydrogen-fueled nuclear fusion reactor for around 60 years and, at best, is still decades away from starting up a commercial power plant with the technology, if it ever happens. By contrast, the cost of developing the LFTR technology and building a commercial-scale power plant with it is estimated to be less than $10 billion. The cost of building an additional commercial-scale PWR power plant with existing technology would be a similar amount. Getting an LFTR built would not require a Manhattan Project–sized or even nuclear submarine–sized effort. With mass production, the cost of an LFTR could be significantly less than a PWR and the required equipment considerably more compact.

According to Bill Gates, lowering the cost of energy and improving its availability would be the best way to improve the lot of the poorest 2 billion people on Earth.[20] If the cost of an LFTR-based power plant were low enough, it could have an unprecedented effect on reducing world poverty. Historically, every time a new energy source has been developed and increased the availability of energy, it has increased overall global wealth. The availability of coal initiated the Industrial Revolution. The availability of oil enabled the creation of modern transportation, everything from the automobile to the airplane. Nuclear power was supposed to be the ultimate source of energy but has fallen short of its promise. The LFTR using thorium could very well be the game changer.

Could Mars be colonized and developed as a spacefaring planet with manufacturing and the ability to make things like rocket fuel, without a thorium-fueled LFTR? Maybe, but without some type of powerful but compact nuclear reactors, energy intensive activities such as interplanetary travel between Earth and Mars as well as asteroid mining would be seriously hampered, thus rendering Mars as nothing more than a scientific outpost similar to Antarctica. The thorium-based LFTR might not be the only possible solution, but it looks like a really good one.

Unquestionably, 100 percent of any nuclear power technology, equipment, and fuel used for powering Mars will come from Earth during the early decades of colonizing Mars. This, more than any other single factor, will be the pressure point Earthlings can use to maintain control over Martians. The day that the first 100 percent Martian-built and -fueled reactor goes on line to power Martian cities will be a necessary milestone in the drive for self-sufficiency, itself a necessary condition for future independence.

NOTES

1. Christianna Reedy, "NASA Wants to Collect Solar Power Directly from Space," accessed June 5, 2019, https://futurism.com/is-space-based-solar-power-the-answer-to-our-energy-problem-on-earth/.
2. John Bluck, "Antarctic/Alaska-like Wind Turbines Could Be Used on Mars," NASA Ames Research Center, accessed June 5, 2019, https://www.nasa.gov/centers/ames/news/releases/2001/01_72AR.html.
3. Leonard David, "Wind-Powered Mars Landers Could Really Work," accessed June 5, 2019, https://www.space.com/41023-mars-wind-power-landers-experiment.html.
4. Bruce Dorminey, "Why Geothermal Energy Will Be Key to Mars Colonization," accessed June 5, 2019, https://www.forbes.com/sites/brucedorminey/2016/09/30/why-geothermal-energy-will-be-key-to-mars-colonization/#7eed970d4b25.
5. World Nuclear Association, "Nuclear Power Reactors," accessed June 5, 2019, http://www.world-nuclear.org/information-library/nuclear-fuel-cycle/nuclear-power-reactors/nuclear-power-reactors.aspx.
6. Ross Pomeroy, "Why Not Nuclear-Powered Aircraft?," accessed June 5, 2019, http://www.realclearscience.com/blog/2014/07/why_not_nuclear-powered_aircraft.html.
7. Dave Mosher, "A Cold War Technology Designed to Make Jets Fly for Days Might Solve Earth's Looming Energy Crisis," accessed June 5, 2019, https://www.businessinsider.com/aircraft-nuclear-propulsion-molten-salt-reactor-2016-12.
8. CNN, "U.S. Nuclear Plant Had Partial Meltdown Years before Three Mile Island," accessed June 5, 2019, http://news.blogs.cnn.com/2011/03/29/u-s-nuclear-plant-had-partial-meltdown-years-before-three-mile-island/.
9. Richard Nixon, "Remarks at the Atomic Energy Commission's Hanford Works near Richland, Washington," accessed June 5, 2019, http://www.presidency.ucsb.edu/ws/index.php?pid=3161.
10. Richard Nixon, "Nixon Phone Call—Nixon Speech on Jobs in California," accessed June 5, 2019, https://www.youtube.com/watch?v=Mj5gFB5kTo4.
11. United States Nuclear Regulatory Commission, accessed June 5, 2019, https://www.nrc.gov/materials/fuel-cycle-fac/ur-enrichment.html.

12. Hilary Huaici Song, "Nuclear Weapons 101: Back to the Basics," Columbia University K1 Project Center for Nuclear Studies, accessed June 5, 2019, https://k1project.columbia.edu/news/nuclear-weapons-101-back-basics.

13. John G. Fuller, *We Almost Lost Detroit* (New York: Reader's Digest Press, 1975).

14. Gil Scott Heron, "We Almost Lost Detroit," accessed June 5, 2019, https://www.youtube.com/watch?v=yotCw66_G1g.

15. Robert Hargraves and Ralph Moir, "Liquid Fuel Nuclear Reactors," Forum on Physics and Society, APS Physics, accessed June 5, 2019, https://www.aps.org/units/fps/newsletters/201101/hargraves.cfm.

16. Robert Hargraves, "Thorium Energy Cheaper Than Coal," Amazon Digital Services LLC, Kindle edition: loc. 3023.

17. Ainissa Ramirez, "Where to Find Rare Earth Elements," Nova Next, accessed June 5, 2019, http://www.pbs.org/wgbh/nova/next/physics/rare-earth-elements-in-cell-phones/.

18. Hargraves, "Thorium Energy Cheaper Than Coal," loc. 2736.

19. James Mahaffey, *Atomic Awakening* (New York: Pegasus Books, 2010), Kindle Edition: loc. 4630.

20. Bill Gates, "Bill Gates Talks about Energy," Wall Street Journal YouTube, accessed June 5, 2019, https://www.youtube.com/watch?v=IsRlN1oDm60.

Chapter Eight

Manufacturing on Mars: A Means of Taming Nature for Survival

Since 1607, when the original inhabitants of Jamestown went ashore to establish the first permanent English colony in America, manufacturing equipment and technology have been the backbone supporting colonial success in distant, isolated, and undeveloped environments. In 1607 the tools required for manufacturing were simple human-powered ones: axes, shovels, hoes, saws, hammers, and so forth. Using these and the abundantly available raw material (trees), the early colonists could build shelters. They could equip their shelters with everything from tableware to furniture, surround their dwellings with fortifications for protection from outside threats, and stockpile firewood for their energy needs.

The firewood was not used just for cooking and heating but also for powering the colony's profit-making ventures. These included refining precious metals (like the Spaniards, they'd hoped to find gold but ultimately failed), smelting copper and producing brass from local sources of metals (again a failure), and fabricating glass objects.[1] A number of Dutch, German, and Polish glassmakers arrived in 1608 with the goal of starting a glassmaking industry. Unfortunately, this too failed, at least in its original effort. It seems that the glassmakers were better at making glass than at choosing friends. They eventually threw in their lot with the local Powhatans and sought to strengthen the relationship by smuggling them weapons, only to be massacred later by their newfound compatriots.[2]

While early firearms and other weapons were not manufacturing tools in the conventional sense, they nonetheless produced two important products: security and food in the form of game animals. The colonists could sometimes trade for food with the Powhatans, but in the end they had to depend largely on the meat provided by firearms and the crops produced using simple farm tools.

The manufacturing technology of the colonists was not just support for potential profit making or colonists' creature comforts. It was vital for survival. The death rate among settlers was horrific with disease, starvation, and attacks by the Powhatans as leading causes (not necessarily in that order). With little to no knowledge about pathogens, let alone medical technology, the ability to create heated shelter was one of the few factors that kept early American colonists from being entirely wiped out by disease. With no fishhooks, firearms, or farm implements, starvation would have been certain. Finally, without the ability to build fortifications and the weapons to defend them—firearms, cannons, pikes, and rapiers along with helmets and breastplates—the colonists would have been quickly slaughtered by the Powhatans, who were not thrilled to have European invaders in their territory.

The ability to make things was critical for maintaining the original manufacturing equipment, but also for creating new equipment. Not only could a broken ax handle be replaced with a newly carved one, but a new ax head could be forged by a blacksmith. For that matter, the forge itself could be built. If kilns were needed for glassmaking or firing pottery, they also could be built. The ability to manufacture gave the colonists a way to expand their capabilities while dealing with unforeseen problems and unexpected situations.

With respect to manufacturing, Martians will be analogous to the early American colonists but with some key differences. The Martians will not have to contend with belligerent locals nor will they be as helpless at fending off diseases. In fact, the only diseases the Martian colonists initially will face will be the ones they bring with them. With proper health screening, these should be minimal.

Due to the sterile environment of Mars, bacteria and fungi that start to grow inside habitats will have little to no competition. This can lead to situations similar to those experienced on the Russian MIR space station (as described in chapter 9), in which mats of bacteria and fungi grew on the walls. Under normal circumstances on Earth, if conditions are right for microbe growth, competition from other microbes will help prevent runaway growth of individual species. On Mars, humidity control using carefully designed air conditioning systems will be the key for preventing possible runaway microbe growth inside human habitats. Unlike air conditioners on Earth, the Martian systems will be set up to collect water removed from the air.

In addition, Martian air conditioning will not just focus on temperature and humidity control, but it essentially will create the air (or at least the oxygen in it) and the air's pressure. Without functioning air conditioning equipment, Martians would be as helpless as premature infants. Maintaining the equipment would be a matter of life and death. To make things worse, ordering even a commonplace replacement part would take months for the delivery.

The Martian food production system is also completely dependent on an array of electromechanical equipment; and although it takes more time to starve than suffocate, food production is, nevertheless, vital to survival. The availability of food would not depend just on air conditioning to provide the proper air pressure and mixture of gasses for growing crops, but possibly also require artificial lighting, and in the case of hydroponics, water pumps. The simple truth is that for humans on Mars, survival would depend on thousands of moving mechanical parts, not to mention thousands of electrical components, continuously functioning.

Survival aside, like the early American colonists, Martians also will be expected to support various wealth-building activities on behalf of their benefactors and sponsors. First, they would be responsible for any maintenance done on spacecraft to be returned to Earth. Next, there would be support for mining activities on asteroids and Mars itself. Finally, there would be scientific endeavors in which equipment had to be built and/or maintained; with some of this equipment functioning as an early warning system for preventing asteroids from impacting with Earth. Stocking spare parts, or in some cases even predicting the design needed for parts, would be problematic at best. Having the ability to manufacture whatever was needed directly on Mars itself would be a key strategy. And having the ability to build new manufacturing equipment with even greater capability would be the key to expansion.

Although there is an abundance of raw materials available on Mars, most would need a considerable amount of information (chemistry), processing equipment, and energy to go from minerals in the ground to useful materials such as plastics, steel, or aluminum. By contrast the trees available to the British colonists in America could be turned into a useful material (lumber) using simple hand tools. A significant portion of materials used on Mars initially would have to be imported from Earth, especially in the early days when Mars was building its manufacturing capability. At that point efficient use of materials would be critical.

Normal machining processes for manufacturing metal objects require that metal be removed from larger pieces in order to make parts. While the removed metal (now in the form of small chips) can be recycled, doing so requires a lot of extra equipment and energy. Normal machining also requires stocking numerous metal bars and pieces in a variety of standardized sizes in the hope that the right size will be available to fabricate the desired part.

The ideal situation would be to deposit metal only in the desired shape rather that remove unwanted metal from an oversized piece. Fortunately, such a system already exists: 3-D metal printing. Although there's more than one way to print metal, typically such a system spreads a thin layer of metal powder on a movable platform inside a chamber filled with an inert gas. A high-powered laser then fuses the powder together in the desired pattern for fabricating the finished part. When a layer is formed, the platform moves down, and the whole process is repeated until the part is complete. Afterwards, the part is removed and heat-treated in a furnace to relieve its internal stresses. The excess powder in the printer is collected and reused for fabricating other parts.[3]

Currently, parts can be made from a variety of metals such as aluminum, stainless steel, and titanium. Due to the layers and slight graininess, parts come out with somewhat rough surfaces that often need to be polished and, in some cases, lightly machined using something like CNC (computer numerical control) equipment to refine the surfaces. When printing in a horizontal direction, the deposited metal sometimes needs to have the printer create vertical supports to keep it from sagging. For example, if the letter T were printed in an upright position, the horizontal part of the T might need to be supported with thin vertical columns. Unlike the excess powder, these vertical supports would be removed later as scrap. Even with final machining, polishing, and removing of supports, there will be less scrap with 3-D printing than with normal machining.

Unlike conventional machining, which requires maintaining an inventory of different sized bars or blocks of metal, 3-D printer powder can be shipped in containers of any size or shape. Hence, having the right-sized piece of raw material for a given job is not an issue. However, at the current selling price of $350 to $450 per pound, the powder is extremely expensive. Shipping to Mars makes the situation even worse.[4] The current printing rate also is slow, and if speeded up too much, the part will sag since lower layers will not have enough time to cool off and solidify before new layers are added. Given that printers are lightweight, mechanically simple devices (as compared to conventional machining tools with numerous moving parts and extremely stout structures), the best way to increase production rate is to use additional printers controlled from the same computer.

Ideally, Martian 3-D printers would be able to increase the number of printers by printing themselves, with the exception of a few parts that might need to be made with other 3-D printers using different materials, such as plastic instead of metal, or with other manufacturing processes. In 2005 Dr. Adrian Bower began pioneering work on a 3-D printer that could replicate itself and demonstrated a prototype in 2008 that could replicate 50 percent of its parts. But the printer only could print plastic parts, not metal or electronic components. The parts also had to be assembled by hand.[5]

Fast-forward a decade, and hype has outpaced actual self-replicating 3-D technology.[6] While contemplating grand schemes of self-replicating 3-D printers that can print not just themselves but also print an entire moon-base—all from materials commonly found on the Moon—researchers at Canada's Carleton University are working mostly on more mundane aspects like printing some of the components, such as the electric motors and electronics needed to control them.[7] Give 3-D printing about a hundred more years of development—about when Mars colonization starts getting serious—and self-replicating 3-D printers will be printing clones of themselves but with Martian materials rather than moondust.

Metal parts made with 3-D printers have tensile strengths (a measure of how hard it is to pull a material apart) similar to pieces made out of the same material using conventional machining processes. But the fatigue strength of metal 3-D printed parts tends to be lower.[8] Fatigue strength is a measure of how well parts resist forming cracks when subjected to repetitive stress cycles. For example, bending and then straightening a piece of wire (like the wire in a paper clip) would be considered a single stress cycle. Do it numerous times and a small crack will form in the area being bent. The crack will then grow with each stress cycle until the wire breaks in half. Small voids or surface imperfections act as stress risers that can greatly reduce the number of cycles required to make a part fail.

Fatigue failures are insidious since the part experiencing them can be used in normal service for long periods of time before developing a crack. If the crack is not immediately detected, it will grow quickly until the part fails catastrophically.

The first commercial jetliner, the de Havilland Comet, which first flew as a commercial airliner in 1952, gave engineers a stunning lesson about fatigue failures. Passengers enthusiastically flew at 35,000 feet in a specially pressurized cabin while viewing the world below through nice-sized square windows. Little did anyone know that the square corners in the windows were stress risers. Each time one of these aircraft flew, its cabin was pressurized when climbing to full altitude, then depressurized before landing—one complete stress cycle. By around 1954, some Comets had accumulated enough stress cycles to cause catastrophic failure of the pressurized cabins. Subsequently, Comets began exploding in mid-flight due to fatigue cracks that had formed at the square corners of the windows (the reason airliner window corners are now always rounded).[9] Needless to say, the reduced fatigue strength of 3-D printed metal parts would be an engineering concern, especially for parts used in critical applications.

3-D printing also can be done with plastics. Again, there are a number of different processes. Some printers use a process similar to the metal printing process described above except with plastic powder. Other printers use a liquid material. In this case, the liquid will polymerize when subjected to certain colors of laser light. Hobbyist 3-D printers typically feed long strands of plastic into heated printer heads, where it is melted and deposited on the printed object. The strands come in a variety of colors and materials that look a little like the plastic cord in string trimmers. As is the case with metal printing, horizontal sections of a printed object need temporary vertical supports to prevent sagging. Again, the current printing rate is slow. It can take several hours to print a single part.

On Earth inexpensive plastic parts are normally injection-molded at very fast production rates in which plastic material is shot into a mold where it rapidly solidifies. The molds are precision-machined and highly polished, with prices ranging from about $1,000 for a mold producing a single plastic washer with each cycle up to $80,000+ for multi-cavity molds producing several complex parts, such as Xbox cases, with each cycle.[10]

3-D printing is far more flexible, requires much less equipment, and currently can make plastic parts in quantities of less than 1,000 cheaper than injection molding can. While a Martian fabrication shop likely will include some conventional machine tools including some form of CNC machine, the predominant manufacturing systems will be 3-D printers. Given that anything on Mars would weigh 62 percent less than on Earth, 3-D printing, whether in plastics or metals, may well be faster and require fewer vertical support elements for preventing sag when printing horizontal structures than the same part printed on Earth.

Although stronger and cheaper, a steel beam on Mars would weigh about as much as a similar sized aluminum beam on Earth, enabling new types of robust construction. Scaling up equipment to larger sizes would be easier than on Earth. Considering that Earthlings already are 3D printing houses out of fast-drying concrete, one can hardly imagine the possibilities for large-scale printing on Mars.[11] With energy supply problems resolved, building massive 3-D printers, capable of making everything from vehicles to other printers, would be conceivable.[12]

Martian Bio-Printing: Beef, Bones, and Pizza

3-D printing is not limited to printing in metals, plastics, and concrete. Research is currently under way to 2-D print—such as printing skin cells on a burn victim—or 3-D print replacement human body parts, everything from bones to functioning human organs like the kidney.[13] The process involves laying down a bio-compatible substrate that acts as a structure for containing living cells "printed" inside it. The cells then are allowed to grow and expand inside the structure. Typically, the cells are cultured ahead of time from the person who eventually will receive the 3-D–printed body part. Unlike today's transplant recipients, who are required to go on immunosuppressing drugs for the rest of their lives to prevent rejection of their new organ, recipients who receive 3-D–printed organs would not need such drugs. Cells in the 3-D–printed organs would come from the recipients themselves, not from organ donors, and there essentially would be no fear of rejection. By the time Mars colonization is well under way, 3-D–printed body parts will be commonly available.

In the interim between 3-D printing laboratory curiosities and printing actual fully approved and functional human replacement parts, 3D printing of various organs, tissues, and tumors holds the promise of eliminating a significant amount of animal testing. It looks like testing drugs or cosmetics on 3-D–printed body parts made from actual human cells may yield more meaningful results than the same type of testing on animals.[14]

The same 3-D printing technology used to print and grow body parts also could be used for printing meat products using muscle cells grown in tanks rather than growing and slaughtering animals. Apparently, this artificial meat tastes and looks like the real thing but is more tender. Currently, elderly people who have difficult chewing and swallowing and possibly animal rights–motivated vegetarians are likely to be the target market.[15] On Mars, which will have no cattle industry, the potential market will be everyone.

Other forms of 3-D–printed foods also may become commonplace on Mars. With a few extra ingredients, 3-D food printing can convert otherwise unappealing foods (spirulina, ground-up insects) into more palatable forms. NASA already has demonstrated the ability to 3-D print a pizza.[16] Unlike the replicator depicted in *Star Trek* movies, 3-D–printed food will not appear out of nothingness. It will require edible inputs, but it will be able to convert them into numerous different types of desirable food products. What's more, 3-D food printers will be able to create all kinds of interesting and unique shapes and designs with the food.

Okay, it would be cold and dusty, but otherwise Mars would be nerd cool and capable of manufacturing just about anything, assuming it could attract the right nerd talent.

The Martian Methane Economy

Pretty much every aspect of Martians' existence, from the ability to immigrate to the ability to supply their every need, not to mention niceties like their ability to breathe, will depend on their chemical industry, and it will be based around oxygen and methane production right from the beginning. Spacecraft, whether they originate from Earth, then travel to Mars, or originate from Mars itself, will need to load rocket fuel produced on Mars before they can leave the planet.

As pointed out by Robert Zubrin's Mars Direct model, lifting the full amount of fuel off the surface of Earth for a complete trip to Mars and back is unrealistic.[17] It would take a rocket far more massive than the ones used in the Apollo moon missions. Furthermore, the payloads with such a system would be too small to support colonization efforts. The answer would be to manufacture fuel (oxygen and methane) on Mars itself for the return trip. Virtually every scheme for colonizing Mars, including the SpaceX scheme, depends on this method, at least to some extent. The system would have to be set up using one-way spacecraft and robotics before the initial human exploration of Mars, let alone colonization, could begin.

The oxygen component for rocket fuel could be produced by removing CO_2 from the Martian atmosphere, pressurizing it, and heating it in the presence of a ceramic catalyst producing the following reaction:

1) $2CO_2 \longrightarrow 2CO + O_2$

The CO or carbon monoxide is toxic and would have to be separated out. In order to be used as a component of rocket fuel, the oxygen then would be liquified and stored in a large tank until loaded into a spacecraft. The same reaction also could be used for producing breathing air in habitats on Mars. Later in the development of the Martian colony, water could be broken into oxygen and hydrogen, assuming that large reliable supplies of liquid water eventually were available.

The otherwise wasted hydrogen from the electrolysis of water could be used to manufacture the other component of rocket fuel, methane, by taking CO_2 from the Martian atmosphere in the following exothermic (meaning it releases heat) reaction:

2) $CO_2 + 4H_2 \longrightarrow CH4 + 2\ H_2O$

The reaction would occur spontaneously when using either a nickel or ruthenium catalyst.[18] In the early days of colonization, liquid hydrogen would need to be imported from Earth as a component for making rocket fuel. Transporting the liquid hydrogen would be conceivable since it contains far less mass than the liquid oxygen or methane used as rocket fuel for the return trip.

In addition to rocket fuel, methane can be turned into a plethora of different useful petrochemical products, including a variety of plastics. Some of these plastics could be used for creating greenhouses.

The simple chemical reactions shown above belie the complexity of creating the chemical plants required to manufacture the products produced by the reactions. Other than in the things they produce, an industrial-sized chemical plant will bear little resemblance to the typical glassware and rubber tubing setups operating at atmospheric pressure in a typical chemistry lab. The plant will require pipes, pumps, pressure vessels, compressors, heat exchangers, and so on. Most of this equipment will run at elevated pressures and need to be made of metal, at the very least steel. If corrosion is an issue, the equipment will need to be made of stainless steel or a more exotic alloy. In the beginning, all the equipment will need to be transported from Earth; but as the colony expands, it will have to be manufactured on Mars, and this will require a steel making industry.

Luckily, not only does Mars have a wealth of iron oxide (responsible for its red color), but the oxygen-making process (see chemical equation 1 above) produces CO or carbon monoxide, which can be used in the following reaction for producing iron:

3) $Fe_2O_3 + 3CO \longrightarrow 2Fe + 3CO_2$

Of course, obtaining iron is just the first step in steelmaking; to finish the process requires carbon and, depending on the desired type of steel, other alloying metals. Fortunately, these materials also are available on Mars. Nevertheless, building the chemical and steelmaking industries to support Martian expansion is going to require a major effort.

To be efficient, the chemical and steelmaking systems need to be large-scale operations. Interesting questions arise about how the equipment could be fabricated on Mars. Could chemical plant equipment be 3-D printed? So far, no one has done anything remotely like it. Chances are that the major equipment manufacturing on Mars for things like large-sized storage tanks and pressure vessels eventually will look a lot like the same kind of manufacturing done on Earth; 3-D printing will be one of the technologies that helps speed up the evolution of Martian manufacturing into heavy-duty capability.

Once enabled by steel production, the Martian chemical industry's critical impact will not be limited just to rocket fuel and producing breathable air in Martian habitats. If supereffective greenhouse gases (SEGGs) are to be massively released into the Martian atmosphere to warm up the planet, numerous remotely operated SEGG manufacturing plants will need to be built and supplied with raw materials.

Due to an initial lack of trees, products that would be made of wood on Earth will need to be made of plastic or possibly a fast-growing woody plant like bamboo. Even bamboo would be grown in a greenhouse covered with plastic film—probably a thermoplastic film such as polyethylene. As mentioned earlier, for common inexpensive thermoplastics (meltable/extrudable plastic types), methane (essentially the same thing as natural gas on Earth) will be the key starting point.

From the standpoint of energy consumption and process simplification, it would be an advantage if usable deposits of natural gas could be found similar to the ones currently exploited on Earth. If Mars was once a wet planet with a relatively thick atmosphere and actual plant life similar to Earth, natural gas deposits also would be likely to exist. The current Martian atmosphere contains traces of methane gas, so while unconfirmed, there is data to suggest that some form of underground natural gas may exist.

Polyethylene is the most common plastic used on Earth for things like garbage bags, food storage containers, bread wrappers, greenhouse covers, and so on. To make polyethylene, methane is first converted to ethylene. The ethylene is then polymerized, which means that individual ethylene molecules are linked together to form long chains usually composed of thousands of individual ethylene molecules, thus forming polyethylene plastic. The "low-pressure" processes for polymerizing ethylene require several hundred psi of pressure (tens of atmospheres) and specialized catalysts while the high-pressure processes require pressures of 15,000 to 45,000 psi (1,000 to 3,000 atm). The two types of processes produce polyethylene with differing properties for different applications. Needless to say, high-quality steel is going to be required for making the equipment for either process in order to withstand the required pressures.

Bio-Development on Mars

Mars itself will be a hotbed of activity for biologists, geneticists, and biomedical engineers, because humans, either deliberately or accidentally, will be creating and shaping their own biosphere. If Martians want to convert their CO_2 atmosphere to one with breathable oxygen, they will need to find or bioengineer organisms that can convert CO_2 into oxygen, such as lichens, cyanobacteria, plants, or who knows what. These organisms will need to survive the initial conditions on Mars long enough to produce the desired level of oxygen in the atmosphere.

The basic laboratory equipment required for doing bio-research will not be massive, hence, will be easily imported from Earth. Combining this equipment with the availability of steel (especially stainless steel), the various forms of 3-D printing, and the substantial amount of Martian bioscience and bioengineering talent will make Mars a productive bio-research/manufacturing center, developing such technologies as growing meat in stainless steel tanks rather than raising animals. The requisite equipment for such outputs will not be as massive as the steelmaking and chemical industry equipment, nor will the processes need to run at high pressures, so a lot of the tanks and other parts could be 3-D printed.

Bio-research/manufacturing also will support the efforts to grow plant crops in greenhouses. Since the Martian atmosphere will contain limited amounts of nitrogen, nitrogen-fixing plants and bacteria may be unable to draw enough nitrogen out of the air to enrich the soil so that plants can prosper in it. Hence, the required nitrogen may need to come from human waste or synthetic nitrogen fertilizers. Fortunately, nitrogen compounds have been detected in the regolith (essentially Martian dirt), which could become a source of raw materials for making synthetic fertilizers.

Ironically, feedstocks of raw materials for crop-enhancing synthetic fertilizers also can be used for making explosives and gunpowder. This dual purpose was discovered by the Germans in World War I (the Haber–Bosch process) and was used for both food and munitions production.[19] These facts were not lost on various world powers, including the United States, which has taken advantage of this wonderfully flexible situation. In peacetime, the nitrate industry helps make food. In wartime, it helps make munitions.

By the time Mars grows to two million or more colonists, it will have the manufacturing capability to arm itself for a war with Earthlings. Considering the flexibility of its manufacturing technology, by historical standards, Martians could do so in record-breaking time.

The taming of nature's resources through manufacturing will happen similar to the way it happened in the Jamestown colony. It will need to happen for continued survival. Otherwise, like the colonists in Virginia, the Martians will face death and uncertainty a long way from their previous home.

NOTES

1. William N. Kelso, *Jamestown the Truth Revealed* (Charlottesville: University of Virginia Press, 2017).
2. Ibid.
3. Alkaios Bournias Varotsis, "Introduction to 3D Metal Printing," accessed June 5, 2019, https://www.3dhubs.com/knowledge-base/introduction-metal-3d-printing.
4. Ibid.
5. Elizabeth Matias and Bharat Rao, "3D Printing: On Its Historical Evolution and the Implications for Business," 2015 Proceedings of PICMET '15: Management of the Technology Age, http://faculty.poly.edu/~brao/3dppicmet.pdf.
6. Amandine Richardot, "Can You 3D Print a 3D Printer?," accessed June 5, 2019, https://www.sculpteo.com/blog/2017/10/24/3d-print-3d-printer/?&&&.
7. Hanna Watkin, "Researchers Developing Self-Replicating 3D Printers to Build Moon Bases," accessed June 5, 2019, https://all3dp.com/carleton-university-researchers-develop-self-replicating-3d-printers-help-build-moon-bases/.
8. Varotsis, "Introduction to 3D Metal Printing."

9. Robert G. Pushkar, “Comet’s Tale,” accessed June 5, 2019, https://www.smithsonianmag.com/history/comets-tale-63573615/.

10. Rex Plastics, “How Much Do Plastic Injection Molds Cost?,” accessed June 5, 2019, https://rexplastics.com/plastic-injection-molds/how-much-do-plastic-injection-molds-costww.

11. Andrea Powell, “How to 3D Print an Entire House in a Single Day,” accessed June 5, 2019, https://www.wired.com/story/icon-house-3d-printer/.

12. Interestingly, the U.S. military has already invested heavily in research and development of 3-D–printed technologies for logistics and rapid fabrication of bridges and spare parts. Gina Harkins, “The Marines Just 3D-Printed an Entire Bridge in California,” accessed June 5, 2019, https://www.military.com/dodbuzz/2019/01/31/marines-just-3d-printed-entire-bridge-california.html.

13. Lucie Gaget, “3D Bioprinting: What Can we Achieve Today with a 3D Bioprinter?,” accessed June 5, 2019, https://www.sculpteo.com/blog/2018/02/21/3d-bioprinting-what-can-we-achieve-today-with-a-3d-bioprinter.

14. Smartech Analysis, “Use of 3D Bioprinting in Drug Discovery and Cosmetics Testing Expected to Reach $500 Million By 2027,” accessed June 6, 2019, https://www.smartechanalysis.com/blog/3d-bioprinting-drug-discovery/.

15. Robert Gorkin and Susan Dodds, “The Ultimate Iron Chef: When 3-D Printers Invade the Kitchen,” accessed June 5, 2019, https://phys.org/news/2013-10-ultimate-iron-chef-d-printers.html.

16. Leanna Garfield, “This Robot Can 3D-print and Bake a Pizza in Six Minutes,” accessed June 5, 2019, https://www.businessinsider.com/beehex-pizza-3d-printer-2017-3.

17. Robert Zubrin, *The Case for Mars* (New York: Touchstone, 1996), p.3.

18. Robert Zubrin, *The Case for Mars* (New York: Touchstone, 1996), p.150.

19. Sarah Menker, “Like Day and Nitrogen: War, Peace, and the Dawn of Fertilizers,” accessed June 5, 2019, https://www.linkedin.com/pulse/like-day-nitrogen-war-peace-dawn-fertilizers-sara-menker/?articleId=8845390223113810724.

Chapter Nine

The One-Way Trip: Who Goes to Mars Stays on Mars

Much like the colonization of the Western Hemisphere, the geographic realities of Martian colonization will have profound personal, political, and economic impacts on the first waves of settlers. While these realities are insufficient for starting conflict by themselves, they could set into motion a chain of events that will play out over several generations with far-reaching consequences for both planets.

The first spaceships going to Mars to start the process of colonization probably will not return. For one thing, they simply will not have fuel for a return trip. They will arrive with the purpose-built robots and equipment required for creating basic infrastructure such as water, power, and rocket fuel manufacturing that eventually will provide the means for return trips to Earth and other ventures such as asteroid mining. In fact, parts from these first spacecraft may very well be cannibalized for creating the various forms of infrastructure, such as shelter or fuel-making equipment, similar to the way Cortez and other conquistadors destroyed their ships upon reaching the New World, partly to use materials from them and partly to ensure that their crew would be making a one-way trip.

When the initial fully automated missions are judged to be successful, they will be followed by landings with a few humans along with more robots and equipment. The goal here will be to complete habitats (possibly inside lava tunnels), greenhouses for food production, 3-D printers for small-parts production, and the chemical processing equipment to produce oxygen for the habitats. By this point, electrical production also may include a small nuclear reactor. Some of the spacecraft will be returned to Earth, but some will be retained on Mars as the nucleus of a Martian space fleet. These would be used for various ventures launched from Mars, including asteroid mining, satellite launching, or long-range transportation on Mars itself, enabling Mars explorers to establish remote outposts away from the more populated areas.

While establishing those outposts might not look like a critical path item for colonization, the legal and territorial implications are considerable. If there are water, mineral, or other resources to be found, the group finding them will have an early claim on exploiting them. Keep in mind that Martians eventually will need to set up whole manufacturing industries using a vast variety of raw materials if they are to become independent from Earth.

Elon Musk has said that anyone buying a passage to Mars on a SpaceX ship would be allowed to return for free,[1] which fits reasonably well with the SpaceX cost structure. From its beginning, SpaceX realized that the capital cost of spacecraft was by far the biggest factor in the expense of going to Mars, far more so than fuel cost. This meant that spacecraft going to Mars must be returned and reused. Taking a few passengers back on the return trip would not greatly impact costs.

Assuming that the free-return policy holds true, no one would be overtly coerced into staying. However, after securing a place in history as a brave pioneer, would someone then want to be branded a quitter? For Martian immigrants, there is also the issue of sunk costs. In getting to Mars they have paid the price of an Ivy League education and gone through months of space travel. After possibly years of trying to make a go of it on Mars, returning to Earth would be a serious letdown—worse than going into debt for an Ivy League degree only to end up taking orders in a fast-food restaurant while residing in the basement of one's parents' house.

There also is the issue of radiation exposure. Unless engineering solutions eventually are discovered to reduce radiation exposure, a one-way trip to Mars may expose travelers to radiation beyond levels currently allowed for astronauts (which allow a 3 percent increase in the risk of getting cancer).[2] A return trip, or for that matter additional trips back and forth, will significantly increase the risks of an early death from cancer or other health problems such as cataracts.

Given the length of the trip, heading back to Earth for emergency surgery, treatment of an otherwise fatal condition, or to attend the funeral of one's relatives is not going to be a realistic option. Not to mention that fleets of spacecraft likely will only be arriving on roughly a two-year cycle. An opportunity for leaving when one is fed up enough to do so may not be immediately available. No matter what the hardships, the social pressures, sunk costs, and delays preventing travel, not to mention the various difficulties in making a months-long space journey, all will favor staying on Mars, even if the return trip is free.

THE SHORT-TERM PHYSIOLOGICAL BARRIERS

Aside from social and psychological barriers, physiological barriers also will discourage colonists from leaving Mars. These include effects of an extended stay in a lower gravity, lower pressure, higher radiation environment, but while we can list them—thanks to research done on astronauts orbiting Earth in the International Space Station (ISS)—we have no research concerning the extended effects on humans of lowering gravity by only 62 percent. We could assume that they are about 40 percent less than the effects from a stay on the ISS, but then, a returning colonist also will probably end up spending a total of about six or more months in outer space making the trip to and from Mars. So, a Martian colonist will be exposed to a mixture of both low gravity and microgravity.

The effects of a stay on the ISS include bone loss, muscle loss, and blood volume loss. With the exception of bone loss, the effects usually can be remedied within about a month of returning to Earth from orbit. Bone loss, however, can take many months and may even be permanent. The problem with making predictions of the lower gravity effects on Martians based on ISS astronaut data is further complicated by the fact that the Martian colonist will be away from Earth's surface for a significantly longer time than ISS astronauts.

In order to reduce the effects of microgravity on board the ISS, American astronauts exercise about two hours a day. Even this amount does not entirely eliminate the effects caused by a lack of normal gravity conditions. Perhaps a similar exercise program could mostly compensate for the lower gravity on Mars. Colonists might need to train for a return trip to Earth as though they were training for a long-distance run combined with a power-lifting competition. On the other hand, there might be long-term effects from the Martian environment that defy simple solutions.

The Long-Term Effects

Genetic effects from time spent on ISS are subtle but nevertheless raise questions about the long-term effects that a much longer stay on Mars might produce. According to NASA, after a year spent on the ISS, astronaut Scott Kelly was still an identical twin with his brother, Mark, who stayed behind. But Scott's genetic expression had changed by about 7 percent.[3] This slightly altered the way his genes react to his environment. The good news is that the effect was relatively minor. The bad news is that the analysis is based on only one year of living onboard the ISS and a single research subject.

Buried within the analysis of the Scott Kelly data is yet another intriguing bit of data: While in space, Kelly's telomeres became significantly longer during his stay on the ISS. Telomeres are like protective endcaps on the ends of DNA strands that help protect the DNA from damage. Shortened telomeres are associated with cellular aging. Indeed, telomere length is thought to represent a type of biological age separate from one's calendar age—the longer the telomeres' length, the younger the biological age. The fact that Kelly's telomeres lengthened on board the ISS suggests that at the cellular level, in at least some sense, he grew younger. How long did it take for Kelly's telomeres to essentially return to normal after returning to Earth? About two days. Would this same effect occur in Martians, and would it have a significant meaning? No one knows, but it has raised some interesting questions.

Depending on how spacecraft and Mars habitats are designed, and on how much time they spend outside their habitats, colonists, especially the early ones, could accumulate significantly more exposure to radiation than an astronaut spending a year on the ISS. This could result in actual genetic damage, leading to a host of potential problems from cancer to infertility or possible birth defects in future offspring.[4] Again, while the effects are thought to be reasonably manageable, no one knows for sure about the long-term risks.

Potential genetic damage is not limited just to human cells. An average person has about as many bacteria cells inside and on the surface as human cells. The human gut alone houses about 40,000 different species of bacteria, with about 100 times as many genes as in the human genome.[5] So there is not just a large number of microbes in a given person, but also a large diversity. This would not be a big deal were it not for the fact that these microbes play an important role in things like digestion and the human immune system. Just by being present, the diversity of microbes found on the skin and in the gut tend to actively compete with unwanted microbes and prevent them from gaining a foothold.

Mutating or killing off a significant part of the normal skin and gut microbes could have a significant impact on health. On Earth this can happen to gut bacteria from the use of antibiotics and can lead to problems such as diarrhea and autoimmune conditions. While getting the correct balance of microbes reestablished in one's gut after they have been disrupted can be difficult, on Earth it is not going to be due to a lack of microbes in the environment. Doing so on Mars is more problematic: the same diversity of microbes is not present.

How the microbes inside Martians are fed is going to be another factor in their types and activity. Long-term residents on Mars would not be exposed just to different environmental conditions but also a different cuisine, which could cause significant differences in their gut bacteria. Although there is as yet no actual data on the subject, Martians could end up with things like strange intestinal disorders if they return to Earth and start partaking of Earth food. However, the problems a former Earthling might experience on returning to Earth from an extended stay on Mars might be minor compared to the effects that a native-born Martian might experience on traveling to Earth.

The Effects of Being Born on Mars

People who emigrate to Mars are most likely to be healthy, young to middle-aged adults, and human nature being what it is, eventually the first of many native Martians will be the result. At that point humanity will to get a profound amount of data concerning the effects of environment versus genetics. For example, there is a considerable amount of speculation that native Martians will be taller than Earthlings. Indeed, Scott Kelly gained about two inches of height during his year on board the ISS. On returning to Earth Kelly's height returned to normal, but the effects might be permanent if one is subjected to lower gravity while growing up. Still, gravity is only one of many different environmental factors that could affect the development of native Martians.

Elon Musk says he will make travel to Mars available to anyone who wants to go, and yet, when the opportunity actually begins to exist, preconditions likely will be added, namely health requirements. Mars is not going to be a good place to relocate if one requires medical treatment. In fact, many human diseases, including minor ones like the common cold, could be eliminated on Mars if would-be immigrants were carefully screened for illnesses ahead of time. Some form of screening is certain to happen. Imagine what it would be like to have a cramped spacecraft loaded with a hundred people all in various stages of diarrhea and vomiting.

International laws have been in place since 1967 requiring quarantine and disinfectant procedures to prevent the undesired transmission of microbes from Earth to other celestial bodies and from other celestial bodies to Earth.[6] Even a spacecraft merely sent into orbit typically has its interior flooded with ethylene oxide and methyl chloride to kill off microbes before being sent into space.[7]

Michael Crichton's 1969 book *The Andromeda Strain* gave a fictional account of an extraterrestrial microbe that ends up wreaking havoc on Earth.[8] The microbe is on a micro-meteor that impacts a satellite, causing it to fall to Earth. The microbe escapes to a nearby town and pretty much instantly kills everyone in it. Fortunately, the microbe eventually mutates into a benign form and floats off into the upper atmosphere where conditions are more to its liking. In 1971, this best-selling book was made into an Academy Award–nominated movie. It has since become a science fiction icon.

Russian cosmonauts onboard the Russian space station MIR were perhaps the first people to directly observe an Andromeda strain–like event, albeit a far less deadly one, when, as noted earlier, they observed a thick fungus mat growing over the outside—yes outside—of a viewing window on their orbiting space station. The fungus—possibly from spores floating in outer space—not only grew rapidly but began eating its way through the titanium quartz window![9]

Over time, the MIR became infected with numerous multicolored mats of bacteria and fungi that began eating away at the interior of the spaceship. Whether these bacteria and fungi were inadvertently brought to the MIR from Earth or from spores in space is subject to debate. Certainly, MIR's problems with air conditioning and humidity control contributed to the fungi/bacteria growth. However, they positively thrived in the high radiation environment and were virtually impervious to attempts at killing them.

NASA tended to downplay the severity of the bacterial-fungal mats until 1997, when American astronaut Jerry Linenger arrived aboard the MIR and observed them firsthand. Linenger was an M.D. and had a doctorate in epidemiology. When he tested the growth rate of bacteria-fungal samples on the MIR, they filled the container and grew over its sides within hours.

Though not as severe as on the MIR, even the newer ISS has problems with rapid bacterial-fungal growth on surfaces. The ISS has more consistent humidity control and more rigorous surface-cleaning procedures. Nevertheless, hanging a sweaty shirt after a workout next to a wall can provide enough moisture to start a bacteria-fungi growth on the wall.

As for space's effect on pathogenic microbes brought from Earth, in 2006 a group of researchers from Arizona State University divided a salmonella culture into two parts and sent one into space on board the Space Shuttle Atlantis. The other remained on Earth growing under identical conditions except for those unique to outer space. When the sample was returned to Earth, the researchers inoculated one group of animals with it and another group of animals with the sample that had remained on Earth. The space sample was nearly three times more likely to cause disease in the test subjects than the Earth sample. It appears that space flight had altered the salmonella's properties for the worse.[10]

Given all the various precautions about the transmission of unwanted microbes by spacecraft, it is unthinkable that infected people will be allowed to travel to Mars. As a result, native Martians are likely to grow up in an environment that is essentially free of Earth's diseases. The downside is that as the native-born population replaces immigrants as the majority, the majority of Martians will have little to no disease resistance to Earth's pathogens.

Martians will not be isolated just from pathogenic microbes; they also will grow up in an environment with a much smaller subset of Earth's microbes in general. As a result, native Martians probably will have significantly different skin and gut microbe profiles than Earthlings. Given that one's microbes can affect everything from emotional well-being to the immune system, the available microbes or lack thereof could have a significant effect on how future Martians look and behave.

The massive number of different mundane microbes on Earth will not be the only biological influences missing. There will be no tick, mosquito, or other bloodsucking, disease-carrying creatures on Mars (unless they are unwittingly carried there by spacecraft). Certainly, some of these animals will lack a suitable habitat and be unable to establish themselves even if they arrive on Mars, but those that do find Mars habitable may produce out-of-control populations thanks to a lack of predators or diseases. Japanese beetles provide a good example. They first showed up in the United States in 1916 in a nursery in New Jersey. While not a significant problem in Japan, in America they have no predators or diseases to control their populations and have become a blight in the eastern United States.[11] "Killer bees" in America can be traced to a single—now much maligned—individual, Warwick Estevam Kerr, who imported 26 queen bees from Africa to Brazil in 1956 as part of a scientific experiment.[12] Unfortunately, his group of queens escaped and spread northward, eventually into the United States, where they have gained a fearful reputation even though the number of people actually killed by them has been very low. Killer bees are able to outcompete honey bees in numerous ways, including having a high immunity to varroa mites, a major honey bee colony killer. Eventually, something undesirable will be transmitted to Mars from Earth. Whether it is deadly or merely an annoyance will not matter. It will trigger a tightening of controls, possibly even an irrational response.

If travel of Earthlings to Mars is subject to restrictive amounts of red tape and regulations designed to prevent the introduction of unwanted microbes, diseases, and critters, travel to Earth by native-born Martians likely will be self-restricted. Martians in this category would be traveling from a fairly benign environment to a virtual petri dish of possible infections and diseases. Plus, there would be exposure to multiple types of pollen with possible allergic reactions that would not be encountered on Mars.

Aside from microbe differences and other Martian environmental effects, Martians are not going to have the same gene pool as Earthlings. Just the hefty price of the ticket to Mars will have an effect on the Martian gene pool, not to mention the fact that future Martian immigrants will need to demonstrate they have the skills to be useful on their new planet: a variety of technical skills such as engineering, computer science, and so forth. Ability to build or fix things also will be in high demand. Multiskilled people will be especially useful. A welder who also can program computers will be more useful than a welding specialist. People with low IQs and/or serious antisocial behaviors are less likely to be able to come up with the price of the ticket and demonstrate the mix of desirable skills needed for employment on Mars. As a group, the original colonists are a lot more likely to look like the chess team than the football team. Early colonists are also likely to have a higher average IQ than the population on Earth from which they came.

While differences in intelligence often can go unnoticed, when they are noticed they can have a negative effect on human interactions. No one likes to think they're stupid, and being in the presence of someone who is significantly more intelligent can make one feel that way. The antidote is to find a weakness, real or imaginary, that counterbalances the intellectual strength. For example, in the digital era, the words nerd and geek impart a degree of respectability for describing a person with computer skills, but applied to males the terms also imply a lack of social skills, girlfriends, or miscellaneous manly virtues. So, while categorizing someone as a nerd indicates that they have a high level of specialize intelligence, it simultaneously implies that such an individual is deficient in other practical skills, hence, are not as street-smart as "normal" people.

Martian cuisine also will have an influence on the health and possibly appearance of native Martians. Cow's milk and indeed most animal-based foods, with the possible exception of fish, will probably not be available or be available only in limited supply or in synthetic form. High fructose corn syrup and refined sugar will be less available, if available at all. Foods created from spirulina or algae may become commonplace along with mushrooms used as a meat substitute, as alluded to earlier. Properly cooked shiitake mushroom can taste remarkably similar to bacon. Lion's mane mushrooms taste like lobster with the added benefit that they can produce nerve growth factors that may be helpful in preventing degenerative brain conditions such as Alzheimer's disease. Certainly, fruits and vegetables will be important. Diet alone may end up alleviating many Earthling ailments including cardiovascular diseases, diabetes, obesity, and so on. Native Martians could end up being a tall, slender, healthy, intelligent, and long-lived variant of humans, but no one knows for sure.

The Effects of Culture, Evolution, and Genetic Engineering

Given a long enough time span, evolutionary influences will shape the nature of native Martians, but those time spans could be thousands if not millions of years. Of course, the time scale for microbes coming from Earth could be a few years, but the effect this would have on Martians is impossible to predict.

Genetic engineering, however, is another matter. By the time Martians begin producing offspring, genetic engineering, not just of plants and animals but of people, will have become a fairly mature technology. If the higher radiation levels of Mars prove to be a problem for food crops, borrowing parts of DNA from the radiation-resistant bacteria or fungi found growing in spacecraft and inserting it in various crop plants would be an option. Possibly the same thing could be done in humans. Diseases and conditions caused by genetic problems in humans could be fixable. Appearance factors like hairiness or facial symmetry could be altered.

Considering its isolation from Earth, Martian culture and language will evolve in significantly different ways than on Earth. Although thick clouds, bodies of water, and forests eventually may exist on Mars after centuries of terraforming, several generations of native Martians will not experience them. In addition, the appearance of the Martian sky is pretty much the exact opposite of Earth's. On Mars the daytime sky is reddish colored while the sunrise and sunset are bluish colored. After a few generations, native Martians will view Earth as a strangely different place.

Due to its lower gravity, Martian sporting events will bear little resemblance to those of Earth. Basketball players should be able to jump as much as 60 percent higher (assuming they have similar mass and musculature structure as Earthlings.) The large buildings and stadiums built on Earth for various sports will be impractical to construct, at least during the early days of colonization when materials are in short supply. There goes sports talk, the heart of much male-bonding conversation. Martians and Earthlings will be following totally different sports, assuming that Martians even have sports to follow. But this is not the only place where there will be disconnects.

Earthling movies, TV, and a vast Internet content will be available to Martians, but by contrast, content available on Earth about Martians will be limited to things like documentaries and YouTube blogs. Once the novelty of humans on Mars wears off, few Earthlings will be paying attention. As depicted in the 1995 movie *Apollo 13*, space travel had gone from the hottest entertainment commodity ever to nothingness, that is until the *Apollo 13* crew were about to die and be lost forever in the darkness of outer space. Mars is simply not going to have the equivalent of Hollywood or major news organizations, at least in its early days. Hence, Martians will know more about Earthling culture than Earthlings know about Martian culture.

The few native Martians who do make the journey to Earth are likely to be viewed as frail (thanks to the difficulty of adjusting to Earth's gravity and microbes), backward (due to the fact that they will have had no exposure to common Earthling experiences), and possibly overly intellectual. They will look different, act differently, and talk differently than Earthlings supposedly with their same language. As a result, Martian visitors probably will give Earthlings an impression of being weird and weak, hence, easily defeated should the need arise.

For Earthlings who travel to Mars, the situation, in some ways, will be worse. Of course, after arrival, it will be immediately obvious that they are not from around here. They are likely to be viewed with a certain amount of trepidation. In some respects, native Martians will be like Native Americans before the arrival of Columbus: They will have little resistance to diseases such as measles. The difference is that unlike pre-Columbian native Americans, native Martians will fully understand how diseases are transmitted. Newly arrived Earthlings not only will be instantly noticeable, but they also will be seen as potential versions of Typhoid Mary, the cook who was a carrier of the typhoid bacterium and allegedly transmitted deadly typhoid fever to numerous individuals in New York City between 1900 and 1907.[13]

If conflict is triggered, the many actual and misconceived differences between Earthlings and Martians will tend to dehumanize the opposing groups and feed aggression toward them. Prior to the World War II Pearl Harbor attack, with a few notable exceptions, the Japanese viewed Americans as a pampered people with very little fighting spirit. Conversely, Americans tended to view the Japanese as racially inferior and technologically inept.[14] Even under the best conditions of direct contact and communication, these real and imaginary differences could create tensions. Unfortunately, the best of conditions will not exist.

NOTES

1. Elon Musk, "Elon Musk Reveals His Plan for Colonizing Mars," accessed June 5, 2019, https://www.youtube.com/watch?v=W9olSzNOh8s.

2. Sarah Scoles, "NASA Likely to Break Radiation Rules to Go to Mars," accessed June 5, 2019, http://www.pbs.org/wgbh/nova/next/space/nasa-mars-radiation-rule/.

3. Erin Brodwin, "NASA Sent Scott Kelly to Space for a Year, and 7% of His Genes are Now Expressed Differently Than Those of His Identical Twin Mark," accessed June 5, 2019, https://www.businessinsider.com/nasa-twin-study-new-results-mark-scott-kelly-2017-10.

4. Jon Rask, M.S., Wenonah Vercoutere, Ph.D., Barbara J. Navarro, and M.S.,Al Krause, "An Interdisciplinary Guide on Radiation and Human Space Flight," NASA, accessed June 5, 2019, https://www.nasa.gov/pdf/284273main_Radiation_HS_Mod1.pdf.

5. Paul Forsythe and Wolfgang A. Kunze, "Voices from Within: Gut Microbes and the CNS," accessed June 5, 2019, http://www.indiana.edu/~abcwest/pmwiki/CAFE/Voices%20from%20within-%20gut%20microbes%20and%20the%20CNS.pdf.

6. Cynthia Wallentine, "How NASA's Planetary Protection Officer Keeps Our Germs from Contaminating Other Planets (& Vice Versa)," accessed June 5, 2019, https://invisiverse.wonderhowto.com/news/prime-directive-nasas-planetary-protection-officer-keeps-our-germs-from-contaminating-other-planets-vice-versa-0176749/.

7. Rhawn Gabriel Joseph, "Space Fungi Are Attacking the Space Stations," accessed June 5, 2019, http://cosmology.com/SpaceFungi.html.

8. Michael Crichton, *The Andromeda Strain* (New York: Vintage Books, 2017).

9. Rhawn Gabriel Joseph, "Space Fungi Are Attacking the Space Stations."

10. Joe Caspermeyer, "Space Flight Shown to Alter Ability of Bacteria to Cause Disease," accessed June 5, 2019, https://biodesign.asu.edu/news/space-flight-shown-alter-ability-bacteria-cause-disease.

11. Jamba Gyeltshen and Amanda Hodges, "Japanese Beetles," accessed June 5, 2019, http://entnemdept.ufl.edu/creatures/orn/beetles/japanese_beetle.htm.

12. Ron Miksha, "The Man Who Made Killer Bees," accessed June 5, 2019, https://badbeekeepingblog.com/2016/09/09/the-man-who-made-killer-bees/.

13. Arminta Wallace, "Mary Mallon, the Irish Woman Who Brought Typhoid to New York," accessed June 5, 2019, https://www.irishtimes.com/life-and-style/abroad/mary-mallon-the-irish-woman-who-brought-typhoid-to-new-york-1.3125437.

14. Simon Worral, "How Racism, Arrogance, and Incompetence Led to Pearl Harbor," accessed June 5, 2019, https://news.nationalgeographic.com/2016/12/countdown-pearl-harbor-attack-twomey-anniversary/.

Chapter Ten

The Inevitable Sources of Conflict: Lies, Misunderstandings, and Taxation Without Representation

Much like the British experience in North America, revolution on Mars is neither preordained nor mono-causal. Many parallels exist between the colonializations of these two different new worlds; hence, it is worth using the rebellion against the Redcoats as a lens to view potential friction and conflict between Earth and the red planet.

HIGHER EXPECTATIONS VERSUS DIFFICULTY OF THE MISSION

Those who made the choice to settle in the New World shared a common motivation—the expectation of something better for themselves and their children. This personal motivation was and is a powerful tool. It gets people to take risks and to sustain the toil and enterprise necessary for building a new society. This almost missionary zeal has been termed the "Protestant work ethic" by Max Weber and provides an intriguing explanation about how many early Americans saw themselves as different and exceptional.[1]

Indeed, given the very real difficulties of establishing a Martian colony, it would be surprising if an analogous "Martian ethic" did not arise out of the early settlers as they worked hard to make their new planet a success. As Martian colonists build their new life on a faraway planet, they would remember their old lives on Earth, but they would focus much of their mental and physical energies on the tasks at hand. In both good times and bad, this common sense of purpose and accomplishment naturally would form common bonds and a sense of community, and Martians would begin to think of themselves as different from and better than those who had remained on Earth.

An exceptional work ethic and expectations for making a better life come with a hidden cost. While feelings of exceptionalism provide hope during hard times and a tangible goal to work toward, they also can become a trap and lead to strife. When high expectations are not being met, humans historically have become frustrated and sought to address their grievances, whether real or perceived.

This tendency for people to become violent when their expectations are thwarted is the key insight for American political scientist Ted Robert Gurr's 1970 classic *Why Men Rebel*.[2] According to Gurr, "Men are quick to aspire beyond their social means and quick to anger when those means prove inadequate, but slow to accept their limitations."[3] Humans expect a better life and become unhappy when they believe that their expectations are not being met.

High expectations are natural but potentially dangerous when they do not match the reality on the ground. Gurr terms this divergence between expectations and realities "relative depression."[4] This relative depression phenomenon has deep roots in human psychology and appears to explain many revolutions in human history.

One of the most interesting findings of Gurr's research is that revolution is not limited to poor or repressed nations. In fact, absolute measures of freedom and wealth appear less important to predicting revolution than the mismatch between expectations and reality among the population. This insight helps explain why the American colonies rebelled against Great Britain despite the fact that they were among the richest group of people in the world and the fact that the average colonist enjoyed rights and opportunities far beyond those available to the average citizen of the mother country.

In the most basic form, the population does not actually need to be suffering to rebel, but they do need to feel as if their aspirations for a better life are not being met—thus Gurr's relative depression![5]

Gurr provides three pathways to relative depression, each of which could impact a Martian colony and lead to revolution. The first is "decremental deprivation," where value expectations remain constant while capabilities fall.[6] This could happen if the Martian colony is not as good as advertised or when there is some major setback or difficulty to the colony. In this scenario, expectations remain high, but the going is tougher than advertised, and people become frustrated even if they are making progress.

Given that the Martian mission will be sold to the first generation as a grand adventure with amazing rewards, the seeds will be sown for future disappointment. The mission will attract fairly young, motivated, and exceptional people who are not used to failures and setbacks. The inability of these successful Earthlings to accept the limitations and difficulties of the Martian colony may result in feelings of anger and despair once they encounter the inevitable setbacks. Even if the colonists are making incredible progress, the reality of Martian settlement is unlikely to match the dreams of these individuals and will lead to a sense of failure and frustration. For this pathway to revolution, the colony need not be a failure but simply more difficult than advertised.

Gurr's second pathway to revolution is termed "aspirational deprivation."[7] In this scenario, value expectations rise while capabilities remain the same. This pathway is also plausible as the initial excitement of landing on a new planet and establishing a colony leads people to double down on their already high expectations. If the first generations work hard but do not see an immediate payoff, these increased expectations may become a trap as their idealism is met with a difficult and less than glamorous reality of hard work with little payoff.

The third path to discontent and revolution is termed "progressive deprivation."[8] Here, expectations grow and capabilities do as well, but capabilities either do not keep up with expectations or begin to fall. In this scenario, successes such as building a sustainable habitat and beginning to develop the local economy encourage the new Martians to increase their already high expectations faster than what is supported by reality.

In a successful Martian colony, this pathway is the most likely to cause frustration and violence as the settlers see many of the fruits of their labors and begin to question why their hard-won successes are not resulting in an even more prosperous daily life. Given that the Martians would remain dependent on a long and expensive logistical support network from Earth, they could easily begin to tell themselves that they would be even more successful if only the Earthlings would give them more resources or freedom to make policy for themselves.

While it is impossible to predict which if any of these pathways to revolution would occur, they are all plausible. Paradoxically, the Martian colony could be an unprecedented success and still foment feelings of discontent if the tangible successes do not match the intangible expectations of the new Martians. Given the incredibly high expectations of those who would choose to travel to Mars, it seems probable that the first generation of settlers would suffer from feelings of relative depression.

These feelings of relative depression may not be as problematic for the children of the original colonists or later generations of settlers. These groups may still think of themselves as different than humans on Earth but may not share the lofty expectations of the first waves of colonists to the red planet. While this may help manage expectations, new problems easily could arise.

Indeed, it would only be natural for the children of Martian settlers to feel trapped on the planet. They did not volunteer for the mission, and they now may be unable to return to Earth for financial and biological reasons. If Martian economic development stalls, the quality of life for these future generations could actually decrease. While they would not have the lofty expectations of their ancestors, decreased opportunities could provide an alternative pathway to relative depression so long as the realities do not meet their expectations.

While rebellion and violence are not preordained, the very fact that Martian colonists may think of themselves as exceptional, combined with the very real challenges to building a successful home, may sow the seeds for discord.

Communications Breakdown

For the reasons stated above, Earthlings should be careful to manage expectations of the Martian colonists. Part of doing so is through active communications between the two populations. Such communications would be essential not only for effectively managing the political and economic situation on the distant colony, but also for maintaining trust and good relations between the two populations.

To return again to the example of the British colonies in the New World, the lack of effective communications led to feelings of separation, mismanagement, and indifference, and ultimately led the colonists to choose their own path. During this period, letters required a minimum of three weeks to travel across the Atlantic under the most favorable conditions, and many took six to eight weeks to reach their final destination.

This cumbersome communications system virtually ensured that neither side could effectively gather relevant news updates about what was happening on the other side of the Atlantic. In the resulting intelligence gap, decisions were frequently overtaken by events, and even well-intended policies were ineffective. This lag in communications was exacerbated by the fact that few decision makers on either side traveled between the two continents and thus were unable to personally assess the situation across the ocean or provide timely advice on the political, economic, and military issues of the moment. Simply stated, ineffective communications led to ineffective policies, and ineffective policies led to conflict.

Despite the advances in technology in the past 250 years, effective communication will remain an issue for a Martian colony. Radio and Internet communications are limited to the speed of light, approximately 300,000 km/sec in a vacuum. When applied to the relatively limited geographical distances on Earth, the time lag in these communications is virtually unnoticeable. Even a single second of delay in a cell phone call is a rare annoyance, and provides only the occasional disruption in effective communication.

When the Apollo astronauts traveled to the moon, their radio transmissions took several seconds to reach Earth, a fact that necessitated stricter radio discipline and occasionally led to garbled transmissions but was ultimately a user-friendly and effective system for relaying messages in (almost) real time.

Extrapolated over the vast distance between Earth and Mars, these physical limitations become a real issue. The distance between the two planets varies greatly, ranging from 78 million kilometers to 378 million kilometers.[9] Signals still travel at the speed of light, but the distance to cover is so great that it results in a lag that ranges from 4.3 minutes to 21 minutes (each way) depending on the relative distance between the two celestial bodies.

While this delay is a major improvement from the age of sail, it still poses a major challenge with real political and social impacts. The grim reality is that even the shortest time lag of 4.3 minutes each way makes having a conversation in real time impossible. Imagine if you sent an introductory greeting when you first started reading this chapter. The signal would probably just now be getting to the other side, and you would just be getting through formal pleasantries by the time you finished this chapter (even assuming the shortest possible distance between the two planets). Anything as simple as planning for who takes out the trash suddenly becomes a time-consuming and exasperating exercise in communication. Tense political negotiations may be frustratingly impersonal and lead to political and social strain.

This difficulty in communication may be alleviated by sending detailed instructions or orders in the form of e-mail messages, but this also incurs significant costs and opportunities for miscommunication and ill feelings. Just as it is impossible to fully interpret subtle things like tone and actual meaning in an e-mail on Earth, serious misunderstandings can occur in written communications sent between planets. They lack the considerable amount of instantaneous feedback from facial expressions and body language one receives while talking during face-to-face communication.

For this simple reason, diplomats and world leaders still prefer to travel and sit down in person to discuss the current problems of the day. But the ability to travel between planets is not an option for Earth-Mars relations. This problem is further exacerbated by the tendency of people to act more aggressively over the Internet than they would in person or over the phone.[10] While trolling on online message boards and in comments sections may provide amusement and frustration for users of online media, this tendency to act more aggressively online may have profound impacts for interplanetary diplomacy. Why not act more aggressively than you would in person via the impersonal medium of e-mail?

The impossibility of having a real-time conversation, combined with the possibility of misinterpreting e-mail communications, invariably will lead to feelings of abandonment, isolation, and political difference. Here again, this is not in and of itself sufficient for starting a conflict, but it certainly does not help the situation.

A well-established literature in the international relations field suggests that conflict is often not a result of deep-seated resentment, intractable differences, or ancient hatreds, but rather a failure to effectively communicate and bargain. When combined with incentives to misrepresent the truth and a belief that other players may be bluffing in their bargaining, it is entirely possible that conflict could arise even among two actors that want peace or could have negotiated a peaceful resolution to their differences.[11] While the theory for miscommunication leading to conflict is elegant, it does require a disagreement or underlying issue to lead to war. Unfortunately, the colonization of Mars will provide many potential areas for disagreement, especially given the lack of effective communication.

The Realities of Racism

These political differences quite possibly could be exacerbated by a new form of racism. As previously discussed, the physical effects of gravity and other environmental factors will lead to noticeable biological divergence between humans on Earth and Mars over time. Martians rapidly will become taller and less muscular than the population on Earth, and their DNA also will begin to evolve on a deviated path.

This is potentially very dangerous. For a complex combination of biological and cultural reasons, human communities develop an almost universal fear of those who look and act different. Hundreds of studies have shown that people fear those who look different than they do and are more willing to use violence on groups that appear dissimilar. This fear of others may have deep roots in our hunter-gatherer ancestors, who had to compete for scarce resources with outside groups and needed a quick means of identifying those who were part of their group and those who were not. In fact, evidence of wars between groups over access to land, food, and other resources appears very common in prehistory, something that stands in stark contrast to the myth of the peaceful or "noble savage." These prehistoric conflicts occasionally resulted in the wholesale slaughter of tribes, a preindustrial genocide that should give modern man a pause for concern.[12]

Take for example the fate of humans' close biological competitor, the Neanderthals. Scientists now believe that despite interbreeding, humans and Neanderthals frequently fought against each other for access to hunting grounds and lands. Over a period of several millennia, humans ruthlessly exterminated the Neanderthal population. While this probably was driven from a desire to survive and propagate their own species, it likely was aided by an innate sense of biological difference. While most people have a deep reversion to killing members of their own species (even in the heat of battle) these niceties did not protect the Neanderthals, who were visibly different from their human enemies.[13] They were killed in part because they were different!

On Earth, this biological instinct helps explain xenophobia and racism, which unfortunately are common across racial and ethnic groups. On Mars, this same xenophobia could lead to mistrust and conflict. Simply stated, it is easier to kill something that looks different than you.[14] While the modern norms of society tell us that racism and violence are bad, it may be difficult to fully ignore our animalistic inclinations. The disturbing fact is that the natural fear of those who look different may make it easier for both sides to kill each other and may create new racial prejudices and fundamental biases, which lessen our reluctance to make peace.

A Generational Problem

As the populations of Earth and Mars grow apart from each other in a literal sense, they also will grow apart from each other in terms of their personal and social ties. While the first Martians will know many people on Earth and may have fond reminiscences of their previous lives on their home planet, future generations will not have actual memories of the home planet. This first generation of Martian natives will still have relatives on Earth and will be told stories of the old world, yet these personal connections will continue to degrade over time. After several generations, the vast majority of native Martians will know no one who lived on Earth. This lack of personal connection will be exacerbated by the extremely limited opportunities to travel back to the mother planet to reestablish these connections.

Over time, these people will form a new identity—that of Martians. At best, connections to Earth will be similar to feelings that many Americans feel about their ancestral home, one of pride, perhaps, but not a tangible association with the past or an opportunity to repatriate to their native lands. While many people in the United States still celebrate St. Patrick's Day, Octoberfest, Cinco de Mayo, and other national holidays, such holidays are often excuses to drink and dress up that have almost no appreciable impact on their national loyalties or identities.

History tells us that this same gradual weakening of social ties and national identity occurred during the American colonial period. While the majority of settlers initially believed they were part of England, over time they began to see themselves as something different. While the Founding Fathers claimed that they were loyal to the crown and wanted to preserve the rights of Englishmen, they were confronted with a reality that was far different. The mother country did not provide for their desires, enforced taxes without their consent, and forced them to carry the burden for their own defense in the French and Indian Wars. Since most of the American revolutionaries had never been to England, and their families and homes were in the colonies, they had little reason to stay once they believed that they were being treated unfairly by an insensitive and repressive regime of which they had little first-hand knowledge or connection.

A similar slow but predictable weakening of social ties could further strain the geographical and biological differences. If the physical appearance of the Martians leads to stereotyping, then it easily could erode any sense of amity that either planet feels for each other. Diaspora communities on Earth often feel a connection to their homelands but are shunned as outsiders when they do attempt to interact with the populations there. For a heartbreaking example, one only has to consider the various back-to-Africa movements in American history. While many well-intentioned people (such as President James Monroe) believed that freed slaves would have a better life in Africa, nearly all of the attempts to repatriate these African Americans to the African continent were dismal failures.

When combined with the other problems of communication and physical distance, it seems probable that the inhabitants of Mars would feel that Mars, not Earth, was their true home. If these loyalties change, there may be little reason to remain loyal to Earth.

Political Issues

According to the political theorist Harrold Lasswell, the core reason for government is to answer the question(s) "who gets what, when, and how?"[15] Governments exist to provide goods and services to their citizens. They do this, in part, by collecting taxes with the expectation that the government will redistribute the money back to the society.

This seems all well and good, but the potential for fraud, abuse, and political upheaval is built into this organizational structure. In every system, some people, the "elites," have a disproportionate influence on the political process and therefore a disproportionate influence over the distribution of goods and services to the "masses."

Explicit in this model is that neither the costs nor benefits of society are distributed equally. In short, some members benefit more and some benefit less—there are winners and losers. To maintain their privileged position to choose how the money is spent and distributed, the elites must skillfully manage the political system and keep the masses happy. They do this through a combination of means including symbols of authority, propaganda, the allocation of goods and services, and the threat of violence to maintain the support of the people. While power brokers have vast resources at their disposal, they must continually manage their government and keep the masses happy.

This "Goldilocks problem" means that elites are always trying to strike the balance that is "just right" for their people. Give them goods and services, but not too much or too little. Demonstrate power and authority, but not so much that it angers the people. Raise money through taxes, but do not wreck the economy or anger the people.

Failure to effectively do so may result in a loss of power and authority. According to Lasswell, history is replete with examples of governments collapsing because the ruling elites do not effectively manage the distribution of power and resources among their populations, or they choose the wrong strategy to sell their policies to the masses.

The most likely path to discord between Earth and Mars is a disagreement about how the elites on Earth are managing who gets what, when, and how on Mars. A myriad of potential political issues could lead to friction between Earth and Mars. Failure to properly resource the Martian colony, an attempt to collect too many taxes from Mars, a breakdown in the health and welfare of the Martian pollution, or simply a feeling that Mars is not being represented in the political decision-making process on Earth all could lead to political friction and ultimately war.

The Outer Space Treaty of 1967 arrogantly, albeit naively, states that "the exploration and use of outer space shall be carried out for the benefit and in the interests of all countries [on Earth] and shall be the province of all mankind." In essence, according to international law, Mars is not for the Martians. To make matters worse, there also will be a conflict between the various factions that want to preserve Mars for scientific study and those that want to terraform and exploit Mars.

Revolutions often have many causes. To return to the case of the American Revolution, it is clear that many factors—divergent expectations, a lack of effective communications, political mismanagement, a shift in identities over several generations—each contributed to the war. After these frustrations reach a tipping point, war becomes a possibility.

Eventually, some issue, or perceived issue, will trigger a flood of built-up negative emotions. Much like the American colonists, the Martians will feel they are not properly represented by their Earth-controlled government and may conclude that separation is the best way to get what they deserve. In Britain's American colonies, the trigger for revolution was a combination of restrictions placed on westward expansion along with ill-conceived taxation. On Mars, the trigger point may well be limitations imposed by Earthlings on further expansion of settlements and terraforming that could destroy valuable scientific information about Mars's past, along with a hypocritical use of Martian space travel capabilities for exploiting mineral resources from both Mars and asteroids for use on Earth.

Once Profitable, Self-Sufficient, and Angry, Why Not Rebel?

Although it is simple in retrospect to claim that the British could have managed their colonies more adroitly and avoided violence, this lacks an understanding of the human condition. People, especially those brave enough to colonize an unexplored world, want a better life and easily can grow frustrated when they do not get what they want. Once the seeds of doubt have been planted, the desires for change and revolution grow.

While the Earthlings would be wise to learn the lessons of history, simply knowing the potential pitfalls may not save them. Expectations are difficult to manage, communications cannot be sent faster than the speed of light, the effects of biological divergence are both inevitable and unknowable, generations will change their identities over time, and political disagreements about the distribution of goods and services are almost unavoidable.

In any given year war is not the most likely outcome, but eventually it will find its way to reach beyond the boundaries of Earth. When war finally breaks out between Mars and Earth, it simultaneously will be frighteningly different in its technological complexity and eerily similar in its human elements.

NOTES

1. Max Weber, *The Protestant Work Ethic and the Spirit of Capitalism: And Other Writings*, edited by Peter Baehr and Gordon C. Welle (New York: Penguin, 2002).
2. Ted Robert Gurr, *Why Men Rebel: Fortieth Anniversary Edition* (Boulder, CO: Paradigm Publishers, 2010).
3. Ibid., p. 58.
4. Ibid., pp. 22–58.
5. Ibid. The trap of rising expectations among democratic citizenry has been an enduring theme in political thought. See generally James Burnham, *Suicide of the West: An Essay on the Meaning and Destiny of Liberalism* (New York: John Day, 1964); John Lukas, *Outgrowing Democracy: A History of the United States in the Twentieth Century* (New York: Doubleday, 1984); and Patrick J. Deneen, *Why Liberalism Failed* (New Haven, CT: Yale University Press, 2018).
6. Gurr, *Why Men Rebel*, 46–50, 60.
7. Ibid., 46, 50–52, 60.
8. Ibid.

9. Richard Boulais, "How Long Does It Take for a Radio Signal to Go from Earth to Mars," accessed June 5, 2019, http://www.physlink.com/Education/AskExperts/ae381.cfm.

10. Aleks Krotoski, *Untangling the Web: What the Internet Is Doing to You* (London: Guardian Books, 2013), esp. 109–21.

11. To name just two of the most prominent examples, see Robert Jervis, *Perception and Misperception in International Politics* (Princeton, NJ: Princeton University Press, 1976); and James D. Fearon, "Rationalist Explanations for War," *International Organization* 49, no. 3 (Summer 1995): 379–414.

12. Lawrence H. Keeley, *War Before Civilization: The Myth of the Peaceful Savage (*Oxford: Oxford University Press, 1997).

13. Dave Grossman, *On Killing: The Psychological Cost of Learning to Kill in War and Society* (New York: Back Bay Books, 2009).

14. One more recent case of racial differences contributing to horrific outcomes was during the Second World War in the Pacific. Here, both sides willing engaged in racial stereotyping of each other, and there is compelling evidence that this contributed to a higher level of brutality than witnessed in Europe. See: John W. Dower, *War Without Mercy: Race and Power in the Pacific War* (New York: Pantheon Books, 1986).

15. Harrold Dwight Lasswell, *Politics: Who Gets What, When, How (*New York: McGraw-Hill, 1936).

Chapter Eleven

Earthlings Prepare to Invade: Going to War by the Public Transportation System

When Germany decided to invade France in 1914 using the Schlieffen Plan, the logistics were daunting. The plan called for moving more than a million troops, their artillery, horses, ammunition, and associated supplies from Germany through neutral Belgium and into France so that they could bypass the French fortifications on French/German border—all to be done before the French could mobilize. Given the conditions of the day's roads and the vehicles traveling them, the only reasonable choice was to use the railroads.[1] In other words, the Germans went to war via the public transportation system with the whole operation depending on careful coordination of train schedules.

The Schlieffen Plan began to fall apart when the Germans arrived in Belgium. Apparently lacking German logic, the Belgians failed to welcome the invaders. Even though totally outgunned and outnumbered, the Belgians decided to fight the opening battle of World War I and in the process, messed up the carefully planned German schedule. To make matters worse, the Belgians destroyed much of their own rail system to prevent German progress. As result, the German army ended up having to walk the rest of the way to France, giving the French time to mobilize and, with help from Great Britain and the United States, eventually defeat the Germans.

Earthlings who decided to invade Mars would face some of the same logistical issues, albeit, in one sense, on a much smaller scale, but in another, on a colossal scale. By the time Earthlings invaded, Mars likely would have at least 2 million people living on it—a number similar to the 2.5 million people living in America at the time of the Revolutionary War. At that war's peak Britain had about 22,000 soldiers in America. In addition, around 30,000 Hessians and 25,000 loyalists served with the British army.[2] But the British were fighting a well-armed public, many of whom had military experience in the previous French and Indian War. Given the Earthling military's modern weaponry and tactics along with the fact that Martians would not be experienced in armed combat, Earthlings probably would need fewer soldiers than the British invaders in America and certainly far fewer than the German invaders in World War I. The distances to travel along with the energy, equipment, and technology required for doing so would be daunting compared to either the Revolutionary War or World War I.

THE INTERPLANETARY TRANSPORTATION SYSTEM

In the early days of exploration, all the resources for the trip to Mars would have been launched from the surface of Earth, but as colonization proceeded, the equivalent of a public transportation system likely would be created to maximize the advantages of the era's technologies as well as the physics-based realities of space travel. As travel from Earth to Mars matured, transportation for the journey could be characterized as three different systems: 1) space shuttles that would transport humans into and back from near-Earth outer space, 2) deep spacecraft that would permanently remain in space carrying travelers from near-Earth to near-Mars outer space, and 3) space shuttles that would carry both people and supplies back and forth from Mars to near-Mars outer space. (As will be explained later, the design of this shuttle might be significantly different than an Earth-based shuttle.) Given the cost of the system and years required to build it, the Earth's military would likely have no choice except to use it for invading Mars.

The near-Earth system would depend on chemical reaction–based fuel for sending a shuttle craft to the deep spacecraft. This would be an expensive operation requiring a large-sized booster rocket in addition to the boosters aboard the shuttle in order to propel it with what could be described as a controlled explosion lasting for several minutes. In other words, this part of the transportation system would have a lot in common with 20th-century rocketry. Chemical reaction–based fuels still would be used because of their ability to violently create massive amounts of thrust in very short periods of time with minimal risk as compared to, say, nuclear reaction–based possibilities. Unlike nuclear-powered devices, in the unlikely event of a crash, chemical reaction–powered craft would present no danger of broadcasting radioactive contaminants. Of course, other options such as a giant aircraft carrying a rocket-propelled spacecraft or a space elevator also might be used but as yet these have not been demonstrated.

When the space shuttles returned to Earth's surface, huge amounts of thrust force delivered by massive multistage rocket-powered boosters would not be necessary. A returning space capsule would need to fire its rocket thrusters in order to slow the capsule's orbital speed so that it could drop out of orbit and fall back to Earth powered primarily by gravity. The capsule could be similar to a 20th-century space shuttle orbiter that would be slowed on its decent into the atmosphere by air resistance and in the final stage fly like a glider and land on a runway. Unlike the 20th-century version, the new one would be operated entirely using computer control without an onboard human pilot. Even if the shuttle did not fly like a glider but only used rocket power and air resistance to slow its decent, the fuel required would be very small compared to fighting against gravity during the launch from Earth's surface. Given these factors, the return trip to Earth would be much less expensive than the outgoing trip, a fact that helps make asteroid mining economical. However, Earthlings invading Mars would mostly be concerned about getting there. To them the cost of the return trip would not be an issue. For one thing, they likely would not return with the same number of people, H-bots, or equipment as they had when they left.

The spacecraft that transported passengers from near-Earth to near-Mars outer space could be a gigantic spacecraft with no resemblance to anything aerodynamic—in some ways similar to the starships of the *Star Trek* franchise. The original *Star Trek* was the first TV show to realize that large spaceships, which usually never landed, were not just a good idea for space-travel but also didn't need to be aerodynamic because there was no air in outer space. The Earth to Mars spaceships would be largely built in outer space using materials from Mars or asteroids. This would avoid the prohibitive costs of launching everything from the surface of Earth. Expecting Martians to supply the materials for constructing a new military spacecraft in outer space that could obviously be used for invading Mars seems like a highly flawed plan compared to simply borrowing the existing public transport system.

Both the spacecraft's main and positioning thrusters would be ion types, most likely using nuclear energy to generate the electrical power needed to ionize the gas and create the intense electrical fields required for propelling the ions out of the thrusters at incredibly high velocities. By using nuclear-powered ion thrusters, the gigantic spacecraft would reduce the main problem of rocketry: the need to expel an enormous amount of mass backwards in order to produce the thrust required to travel forward. The amount of mass that has to be expelled can be reduced with no loss of thrust by increasing its velocity out the back of the rocket, and here, ion thrusters excel. As compared to chemical reaction–based thrusters, ion thrusters are in their infant stage of development, yet their exhaust velocities exceed those of chemical reaction–based thrusters by more than a factor of seven.[3]

Still, for ion thrusters, current performance merely hints at what could be possible after a few more decades of development. According to Einstein's theories, an object's mass will approach infinity as its velocity approaches the speed of light. While no one expects ion thruster exhaust velocity to reach the speed of light, it eventually could be high enough to actually begin experiencing significant relativistic effects. Hence, an ion thruster not only would expel mass at a vastly higher velocity than the chemical reaction–based one but, due to relativistic effects, the mass of the ions would be significantly greater than if expelled at chemical reaction–based velocities. Hence, a nuclear-powered ion thruster would require a fraction of the expelled mass that a chemical reaction–based thruster would need for producing the same thrust impulse.

In addition, for a given mass of fuel, the energy released by a nuclear reaction would be millions of times greater than the energy released by a chemical reaction. For example, a kilogram of thorium used in a breeder reactor will unleash over 10 million times as much energy generated from reacting a kilogram of a hydrogen/oxygen rocket fuel mixture.

Once in outer space, there is no need to produce the brief violent levels of thrust required for blasting off from Earth's surface (the key advantage of chemical reaction–based thrusters) A steady level of thrust over a longer time span works just fine for propelling a spacecraft along its desired trajectory, and this is what ion thrusters are best at providing.

As an added benefit with military applications, some of the electrical energy generated by nuclear power could be stored in super capacitors and used for firing high-energy weapon systems such as lasers or rail guns. While super capacitors cannot hold as much energy as a battery, unlike a battery, they can store it or release it almost instantaneously. Super capacitors are like an electrical catapult spring: The spacecraft still would need to carry projectiles to fire in its rail guns but would not need to carry the gunpowder required for using conventional cannons.

The projectiles themselves would not need to contain explosives. Their kinetic energy alone, a result of their super-high velocities, would give them all the destructive energy required. In fact, this is the basis of the tank-killing sabot round used in the U.S. Abrams tank. When fired, a long, narrow diameter solid dart can be sent hurling toward an enemy tank at a speed of 5,710 ft/s (1740 m/s)—slow by rail gun standards—blasting through armor and creating havoc inside the targeted tank.[4] For civilian space flight purposes, a rail gun could be used to deflect or destroy space junk in the path of the spacecraft. For military missions, such a device could be used for destroying incoming missiles, attacking spacecraft, or early warning space probes designed to warn Martians of an invasion. While such weapons could readily be retrofitted to a civilian spacecraft, they likely would already have them.

The same super capacitor system used to fire a rail gun also could be used for firing high-energy laser pulses. While getting hit by a laser pulse probably would not be as devastating as being hit by a rail gun projectile, it still could be used as a countermeasure against collisions with space junk or in a battle situation, do significant damage to a hostile spacecraft. Laser weapons would not require storing the extra mass of rail gun projectiles, and the velocity of the laser pulses (at the speed of light) would make rail gun projectile speeds look ridiculously slow.

Solving the Microgravity Problem

Considering the serious effects long-term exposure to microgravity has on bones and muscles, the deep spacecraft would need to use one of three possible options: 1) Place the humans in some form of suspended animation; 2) provide sophisticated exercise equipment for simulating the physical stresses normally produced by gravity; or 3) create some form of artificial gravity. Placing humans in suspended animation would significantly reduce their food requirements; and since they would not need to move around, passengers or soldiers could be packed together more closely, thereby making the spacecraft smaller. We somewhat know how to do this. It's called putting a patient in a coma. The problem is that people in comas do not one day wake up, grab their katanas, and start jumping around while chopping body parts off legions of bad guys as depicted in the movie *Kill Bill*, nor are they likely to feel in the mood for an invasion. In fact, they're going to have a hard time standing, let alone jumping around. On the other hand, H-bots could probably be powered down with little effect on later capabilities when full power was eventually restored. Undoubtedly, they would make up at least part of the invasion force.

Sophisticated exercise equipment can reduce some of the effects of long exposure to microgravity in outer space, but the exercise only lasts for a small portion of the day. The rest of the time the space traveler is at the mercy of microgravity, and it causes numerous physiological changes. For example, bodily fluids collect in the face and cause swelling. Even with exercise, sensory systems responsible for balance, stabilized vision, and the perception of simple things like up and down go unused and slowly atrophy. Twentieth-century astronauts tended to have vision, balance, and movement problems immediately after returning to Earth from extended periods in outer space. These are not helpful symptoms for invaders to experience on arrival.

To provide for bodily needs, exercise, and in general, to keep passengers, from going bonkers, each person on board the deep spacecraft would need a significant amount of internal space to move around in. Of course, the military would be far less concerned with creature comfort, so the system originally designed for paid customers could hold many more troops.

A substantial amount of room also would be needed for each person's food, water, and oxygen. To the extent possible these items would be recycled, but this too would require equipment and space. Some portion of the needed food could be grown aboard the spacecraft. By any measure, economically providing all these necessities for people willing to pay for an expensive passage would likely require a very large-size spacecraft. Again, this would enable Earthling generals to pack in more troops since their rations would be minimal compare to paid travelers. As mentioned earlier, H-bots in powered-down form would be a real bargain since they would require no storage for food, water or oxygen.

While the idea of providing a form of artificial gravity sounds beyond any known principle of physics, it's actually fairly simple to do, but somewhat expensive.

When a person on Earth stands on the floor, the downward force of gravity will act on him, but wait—unless a force acts upward, he will sink through the floor. Fortunately, there is a force called the normal force: The force the floor exerts upward on him, the one that prevents him from the embarrassment of rapidly sinking through the floor and subsequently zooming downward toward the center of Earth. Take away the normal force, and he will be in free fall. What would it feel like? Weightlessness (assuming that he could somehow still breathe and stay cool as he falls toward the center of Earth; otherwise, it would feel like dying).

If only one force acts on him pushing upward on his feet, he will accelerate in the direction of the force. If the force constantly points in the same direction, he will fly off in a straight line at an ever-increasing speed. If the acceleration is equal to 1.0 g, he will feel the same as if standing stationary on planet Earth because he has exactly the same normal force. Unfortunately, a linear acceleration of 1.0 g requires a lot of fuel and cannot be maintained for any length of time even remotely approaching that required to reach Mars. The solution is to use a rotating cylinder that converts the linear acceleration to rotational or centripetal acceleration.

In the case of a rotating cylinder, the normal force acting on the person (i.e., centripetal force) is constantly changing direction so that it always points toward the center of rotation, which in turn creates centripetal acceleration. If the centripetal acceleration at the person's feet equals 1.0 g, his feet are going to feel like they are subjected to Earth's gravity. The beauty of this situation is that if the cylinder is located in outer space, once the rotation is up to speed, no further energy input would be needed to maintain it. Presto, we have artificial gravity (actually, artificial normal force, but it feels like gravity).

Many movie depictions of Mars-bound spacecraft, including the fairly accurate 2015 movie *The Martian*, make use of this principle. In his book *The Case for Mars*, Robert Zubrin proposes rotating the crew quarters attached with a long cable to a section of rocket booster that has expended its fuel and is otherwise useless.[5] This would produce the artificial gravity effect (see figure 11.1).

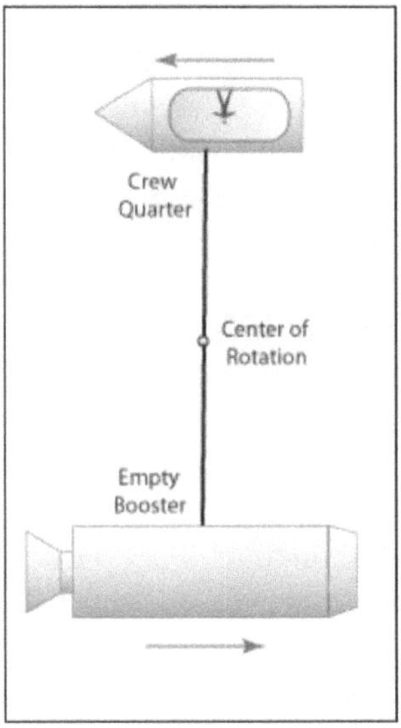

Figure 11.1. Zubrin's proposed solution for creating artificial gravity during a journey to Mars. *Source*: T. K. Rogers.

Of course, no incredibly simple system for producing artificial gravity would be complete without a problem, and this is where the expense comes from. The centripetal acceleration is directly proportional to the distance from the center of rotation. So, if the top of a person's head is at the center of rotation it will have a centripetal acceleration of zero while her feet will have a centripetal acceleration of 1.0 g. If she sits down or stands up from a sitting position, her inner ear will be driven bonkers by the changes in centripetal acceleration. It is likely to be a barf-o-rama. The straightforward but expensive solution: make the cylinder so large in radius that the height of a person is negligible by comparison.

The above is why Zubrin proposed attaching the crew quarters with a long cable to a used rocket booster section and rotating the whole unit rather than simply rotating only the crew quarters about its axis. Zubrin's system would have a very large radius of rotation.

A rotating spacecraft (see figure 11.2) probably would need a radius of around 260 meters (1,706 feet), maybe more, to simulate gravity without producing nausea when people stood up or reclined. In such a spacecraft, when standing, the difference in centripetal acceleration between one's feet and the top of one's head would be less than 1 percent. Admittedly, no one knows for sure whether this small difference would have any long-term effect, but it seems like it wouldn't. Such a craft could have a doughnut-shaped hull and rotate at a little less than 2 RPM in order to simulate Earth's gravity. The circumference would be about a mile (1.6 km), and so the sense of confinement for paid passengers would be minimal. Walking around inside the spacecraft would seem like walking around in a large shopping mall. Getting the spacecraft rotating would take a fair amount of energy, but once accomplished, given the friction-free environment of outer space, maintaining the rotation would require very little effort.

This raises the issue of whether the deep spacecraft should attempt to simulate Earth's gravity or that of Mars. Perhaps it should simulate a value halfway in between. Even better, the deep spacecraft's rotation could be slowly decreased on the trip to Mars and increased on the trip to Earth to acclimate passengers to their next gravity environment. Extra fuel would be required to do so. In any case, having simulated gravity would significantly help keep human soldiers battle-ready during the long journey from Earth to Mars. Although military H-bots would likely not need gravity to stay battle-ready, as a side benefit of the large-sized transport, a very large number of H-bots could be stored aboard in a powered-down condition.

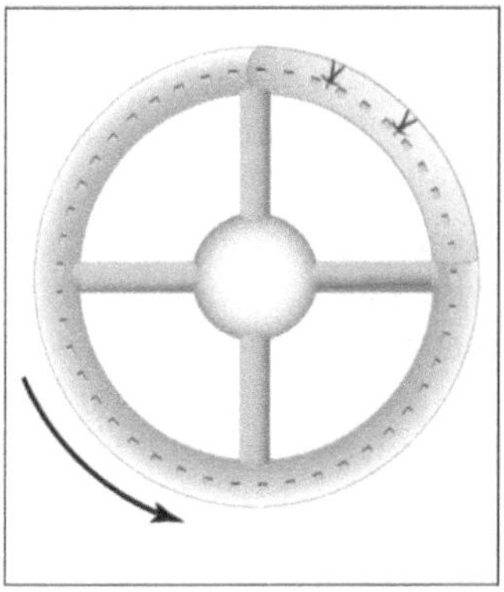

Figure 11.2. Rotating spacecraft that creates artificial gravity. *Source*: T. K. Rogers.

While it might seem like a deep spacecraft of the required size for gravity simulation would be impossibly large, consider that a Nimitz-class aircraft carrier has a length of about 333 meters and is designed to withstand hurricane-force winds and waves. With a crew of around 6,000, such a craft is essentially a small self-contained city, but thanks to its uranium-based nuclear power, one that can travel the oceans for over 20 years without refueling.[6]

The rotating spacecraft also would be nuclear-powered, but likely based on thorium rather than uranium. As explained in earlier chapters, thorium would be a by-product of mining rare-earth metals on asteroids, Earth, or Mars and hence readily available at comparatively low cost. Used in a liquid salts breeder-reactor system, thorium would provide over 100 times as much energy per unit of mass as the uranium-based fuel used in today's nuclear-powered aircraft carriers and in a more compact reactor package. Powering a small, mostly self-contained mobile city with a vast low-cost energy source such as thorium would make the whole system possible.[7]

The Lifeboat/Shuttle Dilemma

Docking a space shuttle with a rotating deep spacecraft would present some design challenges. The deep spacecraft could have a nonrotating space shuttle hangar in the center of the rotating ring, and if it were kept at zero internal pressure, there would be no problems with maintaining leak-free rotating seals between the deep spacecraft's rotating and nonrotating parts. But making the jump from the nonrotating to rotating parts could have its own set of design problems. Having the entire spacecraft, including the space shuttle hangar, in its center rotate is probably a less complicated design.

The 1968 Stanley Kubrick film *2001: A Space Odyssey* contained a scene depicting, in amazingly correct detail, a spacecraft docking with an orbiting space station similar in design to the deep spacecraft described above. In the movie, the docking hangar at the center of the space station rotates with the outer ring. Before entering the hangar, the incoming spacecraft rotates around its longitudinal axis so that it matches the RPM of the space station. Of course, transferring passengers from the hangar near the center of rotation at a centripetal acceleration near zero to the outer ring with a centripetal acceleration of 1g might induce vomiting, but then sometimes a little creature comfort has to be sacrificed for the sake of simplicity of design. For military personnel on their way to war, a little vomiting would be a minor annoyance.

The docking activity raises yet another interesting design consideration. Would the shuttle spacecrafts be based on a home planet and return to it when they had unloaded their passengers and cargo? Or would they remain on the deep spacecraft for the entire journey to the next destination? There they would be offloaded similarly to cargo containers on oceangoing ships, except that they would contain the means for traveling from the deep spacecraft to the planet's surface on their own. This would increase the mass of the deep spacecraft and subsequently increase its fuel consumption; however, the shuttle spacecrafts could also function as lifeboats.

The deep-spacecraft undoubtedly would have highly developed damage control systems and even a certain amount of armor plating to resist damage from collisions with possible micro-meteors or other forms of space junk. The outer shell of the spacecraft perhaps could be self-sealing, similar to the fuel tanks used in World War II aircraft. If punctured by a machine-gun bullet, an elastomeric material inside the fuel tank would stretch as the bullet passed through, then snap back to its original size and seal the hole, thereby preventing an otherwise disastrous discharge of highly flammable fuel. Indeed, given the rigors of outer space travel, civilian-type Mars transport spacecraft would have many features of a warship.

Nevertheless, if a spacecraft were completely disabled in deep space, there would be essentially no chance of rescue. It would be up to the passengers and crew to save themselves. The tiny escape pods typically depicted in various sci-fi movies would be useful only if the disabled spacecraft were orbiting a planet capable of sustaining human life—good luck with that. The choices: Earth and maybe Mars if properly terraformed—in other words, only at the very start and ending of the journey. To be of real value, escape vehicles would need to be able to complete roughly half the journey, maybe more.

Given the massive size of the deep spacecraft, the added mass of the shuttles would likely not make an unacceptably large difference in fuel consumption, but whether they made the entire journey inside the deep spacecraft could depend on design differences between the Earth and Mars shuttles.

Landings could require significantly different shuttle designs, but mainly if the respective shuttles were expected to glide in and make an airplane-like landing. Even a terraformed Mars will never have the same atmospheric pressure as Earth, so the aerodynamics for a glider-based landing could be quite different on the two planets. A landing on Mars likely would require thrusters for slowing the final stages of descent, with the shuttle landing in a vertical position rather than gliding in for an aircraft-like landing. The proposed design for the SpaceX BFR vehicle already solves the problem of differences between landing requirements on the two planets: Both the Earth and Mars landings would be in a vertical position using identical shuttle designs. In both cases, rocket thrust would be used to slow the final descent.[8] Neither landings on Mars nor Earth would be done with gliders. In other words, a single shuttle design would be sufficient.

Having the shuttles make the journey with the deep spacecraft has some engineering issues but seems like the best option, especially for an invasion. Imagine the response Earthlings would get if they had to ask the Martians to send shuttlecraft for offloading their invasion force onto Mars. Besides having multiple shuttles aboard would enable an invading force to land at multiple sites, making it more difficult for defenders to repel the landing.

Dealing with Radiation

Unfortunately, collisions with space junk would not be the only hazard for passengers traveling between Earth and Mars. There is also a health-threatening, maybe even life-threatening increase in exposure to radiation. While soldiers are likely to be less concerned with long-term health effects than civilians, they will not be enthusiastic about being deployed to Mars knowing that the venture will significantly increase their risk of cancer.

The radiation risk comes in several forms: high-energy charged particles from solar flares and cosmic rays, neutron bombardment, and gamma radiation. Most of the radiation comes from charged particles—from the Sun and especially high-energy charged particles from sources outside the solar system.[9] A high-level bombardment of them emitted from solar flares could be predicted by monitoring solar activity. Visible images of the flares will arrive well before the slower-traveling charged particles hit. Deep spacecraft passengers could then presumably take cover in specially shielded compartments.

Extra radiation shielding would be needed on the inner side of the deep spacecraft's outer skin, and this means adding extra mass. More mass means more energy would be needed to create the velocity changes required for the spacecraft to reach its destination. Certainly, new low-mass shielding materials would help. For example, the International Space Station currently uses plastic materials for radiation shielding. A generous supply of energy stored in an extremely compact form such as in thorium instead of rocket fuel could more than offset the increase of mass required for shielding. On the other hand, there is some good news about radiation: It consists of charged particles; and thanks to the charge, there are ways to deflect it.

A coil of wire wrapped around the deep spacecraft's outermost diameter with a current passing through it could produce a magnetic field that would deflect at least some of the high-energy charged particles similar to the way the Earth's magnetic field deflects them in and, hence, protects its inhabitants. The coil would be made of a superconductive material with essentially zero electrical resistance, hence, requiring very little energy to operate. Such superconducting materials need extremely low temperatures to operate, but in outer space this is not a problem.

To deflect charged particles coming from the Sun, the magnetic field would need to be perpendicular to a line drawn between the Sun and the axis the deep spacecraft was rotating around. One way to do so would require the deep spacecraft to move forward like a thrown Frisbee. To visualize the other way the deep-spacecraft could move, imagine the spacecraft were like a doughnut, and we were looking through the doughnut's hole; the spacecraft would be flying in the same direction we were looking. In the second mode of travel, the main thrusters would be aimed in the opposite direction we were looking. Since there is no air resistance in outer space and since the deep spacecraft does not have to land, the craft's forward-facing shape is not an issue. The mode of travel selected would depend on the engineering difficulties involved in how the main thrusters had to be attached. It remains to be seen if the magnetic field and its orientation would adequately deflect charged particles coming from sources outside the solar system.

Other Public Space-Transport Options

As an additional fuel-saving benefit, the deep spacecraft's trajectory could be configured so that it uses the "Aldrin Cycler" system. In this system, named for the person who proposed it, Apollo astronaut Buzz Aldrin, the deep spacecraft would follow a somewhat complex orbital trajectory that would use both the Earth's and Mars's gravitational pulls to help circulate the spacecraft continuously around a path between the two planets.[10] Several deep spacecraft would be spaced along the path and essentially remain permanently in motion back and forth between Earth and Mars. It would be like a railroad system that went around on a strict schedule.

Another possibility would be to modify the orbits of asteroids so that they pass near both Earth and Mars. Assuming they are large enough, the asteroids could be modified over time to form habitats in their interiors. Rotating the modified asteroid could create artificial gravity. Given their considerable mass, the asteroids could easily protect travelers from the ill effects of outer space. While modifications for use as a deep spacecraft might seem nearly impossible, consider the fact that the colonization of Mars likely will occur over at least a few centuries. While it is difficult to predict the exact form, it makes sense that given such a lengthy development and construction time, some form of "public transportation system" for safe and reasonably comfortable travel to Mars eventually would exist by the time Earthlings decided to invade—a transportation system that could successfully be coopted for military use.

In the SpaceX model of transportation to Mars, BFR spacecraft would be sent to near Earth orbit using a massive first stage booster that would detach, fly back to Earth, and make a vertical landing on its launch pad. The first stage booster could be reused, or others like it could then be used to send a different BFR vehicle, configured as a fuel tanker, into space for refueling the original BFR spacecraft. This refueling process would be repeated until the original BFR spacecraft had enough fuel to complete its trip to Mars.[11] While this system lacks the benefits of simulated gravity, it has an advantage for an invasion force: Defenders cannot wipe out the invaders out by destroying a mere handful of spacecrafts.

The cumbersome process of tankers making numerous trips back and forth to fuel the BFR spacecraft so that it could complete its journey from low Earth orbit to Mars could be eliminated by using a tugboat-like booster that remains in outer space. This pilotless computer-controlled booster would use nuclear power and ion propulsion. It would connect itself to the BFR spacecraft while in near-Earth orbit and provide the thrust needed to push the BFR spacecraft to near-Mars orbit. At that point the tugboat booster would detach and the BFR spacecraft make its landing on Mars. The tugboat booster would remain in near Mars orbit until being connected to another BFR spacecraft sent from Mars's surface. The tugboat-booster would then push the BFR spacecraft from near-Mars to near-Earth orbit and then repeat the whole process.

Whether they used the public transportation system, built their own dedicated military system, or used a combination of the two, the Earthling military would need to leave heavy equipment like tanks and artillery at home, but given modern handheld weapons used against mostly unarmed and inexperienced defenders, why would the invaders care?

On the downside, the problem of using the public transportation system for an invasion is obvious—it is public. Hence, any Martian sympathizer on Earth could potentially detect and report it, thereby giving the Martians ample time to prepare.

NOTES

1. Amanda Onion, "Was Germany Doomed in World War I by the Schlieffen Plan?," accessed June 5, 2019, https://www.history.com/news/was-germany-doomed-in-world-war-i-by-the-schlieffen-plan.
2. American Battlefield Trust, "American Revolution—FAQs," accessed June 5, 2019, https://www.civilwar.org/learn/articles/american-revolution-faqs.
3. NASA, "Ion Propulsion," accessed June 5, 2019, https://solarsystem.nasa.gov/missions/dawn/technology/ion-propulsion/.
4. General Dynamics, "120mm KE-W A1® APFSDS-T," accessed June 5, 2019, https://www.gd-ots.com/munitions/large-caliber-ammunition/120mm-kew-a1/.
5. Robert Zubrin, *The Case for Mars* (New York: Touchstone, 1997), 125.
6. Shamseer Mambra, "USS Nimitz: One of The Biggest War Ships in the World," accessed June 5, 2019, https://www.marineinsight.com/types-of-ships/uss-nimitz-one-of-the-biggest-war-ships-in-the-world/.
7. Elizabeth Svoboda, "The Truth about Thorium and Nuclear Power," accessed June 5, 2019, http://www.popularmechanics.com/science/energy/a6162/the-truth-about-thorium-and-nuclear-power/.
8. SpaceX, "Making Life Multiplanetary," accessed June 5, 2019, https://www.spacex.com/mars.
9. Sarah Scoles, "NASA Likely to Break Radiation Rules to Go to Mars," accessed June 5, 2019, http://www.pbs.org/wgbh/nova/next/space/nasa-mars-radiation-rule/.
10. Buzz Aldrin, "Aldrin Mars Cycler," accessed June 5, 2019, https://buzzaldrin.com/space-vision/rocket_science/aldrin-mars-cycler/.
11. SpaceX, "Making Life Multiplanetary."

Chapter Twelve

Collaborators, Spies, and Panic: Nothing Focuses the Mind Like the Hangman's Noose

A war between Earth and Mars would be an unusual mixture of the ancient and modern. While technologies such as space transport, H-bots, and advanced weapon systems would alter much of the tactical character of the conflict, the ultimate outcome of the fighting would be determined by the loyalties, attitudes, and political will of the respective populations.

For a combination of factors, the most likely outcome will be a Martian insurgency in which Earth struggles to reassert control and authority over the Martian population. This type of warfare would be protracted and messy, "like eating soup with a knife," and would be very difficult for Earth to win.[1]

WHY MASS DESTRUCTION IS NOT A VIABLE OPTION

While either planet could simply wipe out the other with some combination of nuclear weapons and kinetic devices (such as large asteroids, rocks, or metal slugs), a rough balance of terror combined with humanitarian (or Martianitarian) and economic considerations probably will deter both sides from using weapons of mass destruction (WMD).

Simply stated, Earth will not want to destroy its investment in Martian colonization, nor will it be willing to pay the political cost of committing the universe's first genocide. Moreover, if plans for an obliteration of Mars leak out, the Martians could preempt Earth and escalate to use WMD first.[2] This would allow both sides to fight, but within acceptable limits, much like the United States and Soviet Union during the Cold War. While both sides would fight to win, they would be unwilling to risk their own annihilation, and a form of cooperation would develop even during a period of conflict.[3]

Assuming that neither side is simply being vindictive, both probably would conclude that completely eliminating the other side is not in their best interest as it would incur unacceptable political and economic costs. If this deterrence holds, and the fear of mutual destruction keeps both sides from destroying each other, strategists on both planets will be forced to consider less destructive strategic options. Rather than a high-intensity conventional war, an insurgency would be the most likely outcome.

A Martian Insurgency

Insurgencies, or people's wars, are among the oldest and most common form of warfare.[4] The conflicts are contests for legitimacy and control of the population. Rather than relying on firepower to kill the Martians, Earthlings would be forced to target key leaders and provide support for those who can be won over to Earth's side.[5]

Paradoxically, Earthlings would be further constrained in their use of overwhelming force, because killing large numbers of Martians would undermine the Earthlings' political support and would drive those on the fence over to the insurgent side. In short, much of the fighting will be a contest fought in, among, and by the people of Mars using relatively light weaponry. In this type of conflict, the people and their "hearts and minds" would be the prize both sides fight to win.[6]

The unwillingness of either side to use mass-casualty weapons and the desire to engage with the Martian people would prove an enormous benefit for the rebellion. Rather than have to fight force-on-force with the more numerous and better armed Earthlings, the Martians will only have to defend their own planet from Earth and keep the spirit of rebellion alive. Earth, on the other hand, will be forced to wage war over millions of miles of space, provide logistical support for their forces, maintain the support of their own political base, and break the rebellion.

As time goes on and costs rise, the insurgency could gain support and legitimacy. By showing their inability to gain control of the rebellious planet, Earthlings will be playing into the hands of the Martians by demonstrating that they are unable to exert control over their colony despite the vast military and economic advantages they enjoy. This will strengthen the political narrative of the Martians, and humans on both Earth and Mars would begin to believe that Martian independence is inevitable. To counter eroding popular support, Earth must win quickly, decisively, and at an acceptable cost in terms of lives and money.

Just as time is typically on the side of the insurgent forces in conflicts on Earth, Mars will be able to win by not losing. Earth must push the pace of the conflict to a rapid termination. For the geographical and physical reasons already stated, this is a very difficult task. The same difficulties of distance, communication, and command and control that precipitated the uprising in the first place will impose the same operational difficulties on the Earthlings.

Further exacerbating this dynamic will be that the Martian insurgents can blend into the local population. Earthlings will be aliens who do not know the local terrain and customs, will look different, have dissimilar clothing and equipment, and will be prone to remain in their bases when not on patrol. On the other hand, the Martians will have local knowledge and contacts, will blend into the local population, and can melt back into it when they choose.

To paraphrase Mao's insurgency doctrine, the Martian insurgents will be able "swim in the population like a fish swims in the sea."[7] This will make it very difficult for the Earthlings to find an enemy to fight and will allow the Martian insurgents to dictate the time and place of fighting on their own terms. In sports terms, Mars will play a home game and can settle for a tie, whereas Earth will have to win on the road in front of a potentially hostile crowd.

While the record of insurgencies suggests that they are only successful about half the time, the unique character of a Martian insurgency tilts the odds heavily in the favor of Mars. Despite the difficulty of the counterinsurgency mission, Earth would feel compelled to reassert control over its colony. Earth forces would be frustrated in their attempts to covertly sneak up on the public transportation system and rapidly capture the entire planet, and a bloody and protracted campaign would ultimately ensue. Unfortunately for Earth, they could quickly fall into a number of traps that would seriously compromise their mission.

Loyalists and H-bots Will Not Win the War

For the reasons stated above, the Earthlings would be powerfully incentivized to win quickly and cheaply. An attractive first step would be to use the people and material already in the Martian theater to kill or capture the Martian rebels. If they could do this, they would avoid having to raise an assault force, send it into space, and then wait the two years for it to reach Mars and begin the kinetic phase of the mission.

While there would no doubt be Earth-sympathetic people on Mars, they probably would be ill-suited to fight against the Martian insurgency. The most obvious problem with using the sympathizers would be their willingness to fight in the first place. In the uncertain early days of the rebellion, these loyalists might not be willing to expose themselves and fight for Earth, given the difficulty of communication and logistical support from the home planet.

Earth's sympathizers would be painfully aware that the planet they were loyal to was a long way off and may calculate that supporting this distant government would be a risky endeavor with an unclear future payoff.

The loyalists also would have to provide for their own operational security in an environment teeming with dissent and rebellion. Since the loyalists would have to consider the possibility that the Martian rebels had penetrated their groups, they would have to act carefully in order not to be compromised and easily killed or captured by the rebel forces. If the Martians could penetrate the inner workings of the loyalist factions, they could easily gather intelligence on their plans and short-circuit any attempt to start a counterrevolution by gaining intelligence on the identities of the loyalists and Earth's plans to employ them.

Even if Earth could identify loyalists, organize them, and induce them to fight without being compromised by Martian counterintelligence, arming them would prove difficult. Logic would dictate that prior to declaring their hostile intentions, the Martian rebels would take the precaution of securing access to weapons and protecting their own bases from attack.

For these practical reasons, even a large group of loyalists may be unwilling or unable to fight for Earth. Given this likely lack of weapons and an effective organizational structure, the best they could hope for would be passive resistance to the Martian rebels until the time when Earth could send them additional military and political support.

Another attractive option might be to try to send signals to the H-bots already on the planet to begin targeting rebel leaders and securing key positions. This would have the additional advantage of striking quickly and avoiding the long journey to Mars, but such a contingency is so obvious that the Martians would likely anticipate the move. To subvert this plan, all the insurgents would have to do would be to disconnect the H-bots from the Earth's network or reprogram them to gather signals from Earth and pass them on without acting on their instructions from Earth. Since the Martian population would be intimately familiar with the design and maintenance of H-bots, this would be a relatively simple task and would again block Earth's attempt to win quickly and cheaply with assets already in the Martian theater of operations.

Earth Gets Serious and Commits Troops to Mars

Given this lack of available fighters on Mars, Earth would have to accept the fact that the conflict required a significant troop presence on Martian soil. Given this reality, Earth likely would choose to send a mixed invasion force of humans and robots to invade and pacify the red planet. While it is possible that such a force already may have been in transit to Mars prior to the outbreak of war, this best-case scenario will still result in a difficult and protracted occupation that will play out over many years and have uncertain prospects for success.

The distance from Earth to Mars would be an enormous advantage to the defenders. In the uneasy "phony war" period between the outbreak of hostilities and the invasion from Earth, the Martians would do their very best to make their defenses as strong as possible to resist the inevitable assault. The time lag also would benefit the rebels by allowing them to root out potential traitors, build their government, provide services for the local population, and reinforce the narrative that they, not Earth, were the legitimate rulers of the red planet.

Once they finally landed, the goal for the Earthlings would be to kill or capture the leaders of the rebellion as quickly as possible, protect critical infrastructure, secure any WMD, and win the hearts and minds of the rest of the Martian population. To be successful, they would need to do this quickly, minimize collateral damage, and convince the undecided Martians that the Earthlings are the legitimate and benevolent rulers of the planet.

This is a very tall order for the Earth force, and failure to achieve any single part of this multifaceted mission could result in either defeat or a Pyrrhic victory. What is more likely is that Earth will attempt to pacify Mars in stages, using an economy of force at first and gradually escalating the conflict. Ships with reinforcements arriving several months apart would mean that a second or third wave of invaders from Earth would arrive only after the initial waves had tried and failed to end the rebellion. While Earth conceivably could send so many troops that it could control all of the Martian landscape and population, the costs of this strategy would require political unity on Earth and a financial outlay that would take decades to recoup.

Sympathizers on Earth

In both detecting the Earthling invasion force and winning the political narrative, the Martians will have a surprising ally in their cause—sympathizers on Earth. Since many aspiring Martians would never be able to emigrate—too old, too sick, too poor—Earth would be a hotbed of sympathizers, collaborators, spies, and political pragmatists. While they may have a mixture of motives for supporting the Martian rebels, they will have the effect of undermining Earth's counterinsurgency efforts and aiding the rebellion. Since Earth will need to commit to a long and expensive campaign against Mars, any discord at home will undermine the political will to see the war to a successful termination.

During the American Revolution, England was no different. Prominent politicians such as Edmund Burke and William Pitt advocated for reconciliation with the colonies prior to the outbreak of hostilities. While their pleas were ignored, they continued to serve as a loyal opposition within the British government, and they slowly began to win others over to their way of thinking. As the costs of continuing the war escalated, both the public and the elites began to see the war as a strategic blunder and began to look for ways to end the conflict with honor.[8]

While the crown could simply have ignored this internal dissent, it ultimately made a decision to yield and end the unpopular and unproductive war, even though it had the economic and military resources to continue. Here again, the American rebels won by playing for time and allowing the political support for the war to erode from within. Ultimately, the British reluctantly decided to let the American colonies go and refocused their energies in the Western Hemisphere on their other colonial possessions in Canada and the Caribbean.

Even if Earth were ruled by a powerful group of elites, private corporations, or a dictatorship, it would not be immune to internal dissent. Like many counterinsurgencies in the past, this erosion of political support and consensus will degrade the willingness to sacrifice blood and treasure in distant lands. Even nonviolent resistance or passivism has its costs, and it is unlikely that there would not be some discord on Earth, if not outright disloyalty or subversion. Over time, internal divisions would force leaders to consider the option of simply cutting their losses on Mars.

Population Control

Perhaps the most effective strategy for Earth to employ would be population control. In such a case, Earth would use its troops to round up Martian civilians and separate them from the Martian insurgent fighters by placing them into camps or controlled areas. Any person found outside the controlled areas would be considered a combatant and subject to imprisonment or summary execution. This would have the advantage of forcing many of the Martian rebels to surrender in order to protect their families and would make it easier to kill or capture fighters, or suspected fighters, outside the controlled areas.

This brutal policy would be difficult to implement but has roots in history. As described earlier, during the Boer War, the British used so-called concentration camps to separate Boer civilians from Boer insurgents. Ultimately, the insurgents realized they could not protect their families and that they could be easily targeted by the British Army if they were found outside the controlled areas.

While effective, this strategy is not without costs. On a humanitarian level, placing people in camps, running the camps, keeping them sanitary and orderly is a difficult task fraught with potential pitfalls and abuses. Even a good-faith effort to run such camps in a humane manner is likely to provoke backlash and resistance on both Mars and Earth.

In the not-too-distant future, even more nefarious and subtle population control measures could be at Earth's disposal. By adopting currently available monitoring devices, GSP trackers, and small computer chips, it already is possible to monitor the activities of individuals on Earth. These technologies could simply be adopted for the more ominous purpose of mass population control of an unfriendly group of rebels or potential sympathizers. This would eliminate the need for traditional camps with guards and barbed wire but would have all the benefits of keeping the Martian population in line. If you resist or do not follow the orders of the Earthlings, you will be captured and fitted with a device that will either shock your body in a painful manner or actually reprogram your brain to cease resisting. While potentially more effective than traditional camps, such a strategy would have significant political costs and have the unintended effect of aiding the rebellion.

Such a backlash might embolden the Martian rebels and accelerate the discord and resistance on Earth. For this reason, population control strategies have been rare during the past century on Earth and may not be politically palatable in the distant future. In sum, even if Earth does adopt this strategy, it is likely that it will suffer defeat in the first space insurgency for political, economic, and military reasons.

Space Is a Vacuum, but Space Wars Are *not* Fought in a Political Vacuum

In thinking about a conflict between Earth and Mars, it is essential to remember that none of these events happen in isolation. Such a war would be fought under intense political, military, and economic pressure. These limits will shape the conduct of the war and limit the freedom of action for both sides. Any slippage in Earth's willingness or ability to fight strengthens the cause of the Martian resistance and makes reasserting control more difficult. Given these difficulties, Earth has three basic options to win quickly and reassert control over Mars: mass destruction, winning hearts and minds, or population control.

To have any chance at controlling Mars, Earth must first get its forces to the red planet, a task that could prove exceptionally difficult and dangerous.

NOTES

1. On this point, see T. E. Lawrence, *Seven Pillars of Wisdom* (Ware, Hertfordshire: Bibliophile Books, 1997), 182; and John A. Nagl, *Learning to Eat Soup with a Knife: Counterinsurgency Lessons from Malaya and Vietnam* (Chicago: University of Chicago Press, 2005).

2. On the importance of keeping escalations covert, see: Austin Carson, "Facing Off and Saving Face: Covert Intervention and Escalation Management in the Korean War" *International Organization* 70, no. 1 (December 2016): 103–31.

3. Jeffrey W. Legro, *Cooperation under Fire: Anglo-German Restraint during World War II* (Ithaca, NY: Cornell University Press, 1995).

4. Max Boot, *Invisible Armies: An Epic History of Guerrilla Warfare from Ancient Times to the Present* (New York: W. W. Norton, 2013).

5. David Galula, *Counterinsurgency Warfare: Theory and Practice* (New York: Fredrick A. Praeger, 1964), 7–8.

6. Mao Tse-tung, *On Guerilla Warfare*, translated by Samuel B. Griffith II (Champaign: University of Illinois Press, 2000).

7. Ibid. p. 93.

8. Peter Stanlis, ed., *Edmund Burke: Selected Writings and Speeches* (New York: Routledge, 2017), 117–252.

Chapter Thirteen

Hollywood Strikes Back: Why the World War II Aircraft Carrier Model Doesn't Work for Outer Space Battles

It's hard to imagine the desperation of U.S. forces in early November 1942 during the battle for Guadalcanal. The navy had only one operational aircraft carrier in the entire Pacific (the USS *Enterprise*). It could launch its planes just fine but not effectively land and service them due to damage inflicted on its forward elevator by a Japanese bomb in an earlier engagement. Normally, landing aircraft would touch down on the back of the carrier, roll forward, then ride the elevator down to a lower deck, where they would be prepared for their next mission. However, given the elevator problem, the aircraft had to land and be serviced at Henderson Airfield on Guadalcanal, an airstrip the Japanese navy was bombarding and Japanese infantry attacking at every possible opportunity. To make matters worse, roughly half the airfield's U.S. Marine Corps defenders were wounded or seriously ill with tropical diseases.

In an attempt to end the battle, the Japanese sent a convoy with 11 transport ships loaded with fresh troops and supplies.[1] Obviously, for the Americans, it would be much easier to destroy 11 large targets in route (the transports) than attempting to destroy thousands of individual targets (the Japanese soldiers) hidden in the jungle. Destroying the transports required striking at a considerable distance with a weapons-delivery system that was much faster (aircraft) than the slow-moving ships. The weapons had to be capable of one-shot target kills (bombs and torpedoes) and have a system for intelligently guiding the weapons to the targets (aircraft pilots). Since the Americans could meet these requirements, they were able to destroy the transports and subsequently cause the Japanese to withdraw from Guadalcanal—a major turning point in the war.

The Japanese transports had no armor (ability to receive weapon strikes with minimal damage) and little ability to destroy incoming weapon systems (antiaircraft capability). In addition, they had no stealth capability. The ships moved so slowly that they could not transport and land their cargoes entirely during a single night, a tactic that could have concealed their activity. During the daytime in open water with clear weather, it was a slaughter.

The transports were accompanied and "protected" by well-armed Japanese destroyers, but given the ineffectiveness of shipboard antiaircraft guns, the destroyers were unable to repel the smaller, more numerous American warplanes. Not only could the warplanes move over ten times faster than the ships, but they also could maneuver in three, rather than two, dimensions. As targets, the aircraft were less than one-tenth the size of most of the ships they attacked. In World War II the only weapon system that had a realistic hope of effectively repelling attacking aircraft were other aircraft—in other words, fighter planes. While the Japanese transports did have some fighters available for air defense, they were nowhere near enough. In World War II, even aircraft carriers defended by their own fighter planes were at times vulnerable to air attack.

When the Earthlings become disgruntled with the Martian insurrection and decide to invade, they are likely to face some of the same issues as the Japanese did when attempting to reinforce their Guadalcanal army. Earthlings are not going to use a death star on Mars. Destroying the colony would be tantamount to destroying trillions of dollars of investment. Earthlings will send a mixed force of robots and human soldiers to occupy the colony and crush the rebellion. Damage the Martian infrastructure too much, and the invaders will not just lack quarters for sleeping; they may very well lack oxygen for breathing. While the invaders will not be using WMD, they will be using transport ships.

At first glance, the idea of Earth sending troop transports through vast distances of outer space to invade Mars almost seems futile, possibly even suicidal, unless the transports are accompanied by the space-travel equivalent of World War II aircraft carriers with fighter-type spacecraft to ward off similar hostile spacecraft. Certainly, the experience of Guadalcanal seems to provide a historic warning.

THE WORLD WAR II AIRCRAFT CARRIER–TYPE SPACE-BATTLE MODEL

Most spacecraft battle images held by the general public invariably come from movies with space fights based on World War II naval engagements in the Pacific. *Star Wars*, for example, typically uses a World War II aircraft carrier model in which swarms of small manned spacecraft attack huge slower-moving vessels at close range—typically no more than a few miles away. In both World War II and the space movies, large vessels are defended by relatively ineffective antiaircraft type weapons. In World War II, gunners lobbed more than 1,000 five-inch standard-issue cannon shells at incoming kamikaze aircraft for every one downed.[2] The gunners defending large spacecraft in movies seem to fare no better.

On the other hand, the Empire's convoys in the *Star War* movies also were protected by their own equivalent of World War II aircraft carriers. These provided swarms of fighter-type spacecraft that made really cool sounds (never mind that there is no audible sound in outer space) and nearly wiped out the attacking rebel good guys (a fine way to build dramatic tension). But as was often the case in World War II carrier battles, in the end, even the fighter defense was not enough to ward off the attackers.

Unfortunately, human pilots guiding small attacking spacecraft only can handle a maximum acceleration of about eight g. To simplify the physics, we'll consider the larger ship to be stationary with the small attacking ship flying around it. At rather sedate, by spacecraft standards, speeds of 1,000 mph (1610 km/hr), an eight g acceleration would limit the small spacecraft to a turning radius of 1.58 miles, not an impressive level of maneuverability when one is at most only a few miles away. Hence, close-range attacks involving a considerable amount of maneuvering likely would occur at speeds below 1,000 mph. By comparison, World War II naval aircraft typically fought at speeds of around 200 to 300 mph.

If the anti-spacecraft weapons fired projectiles, they would fly in a straight line with essentially an unlimited range. By contrast, World War II projectiles would be slowed by air resistance and pulled downward by gravity, making them travel in parabolic arcs that limited their range to a few miles. Thanks to superior technologies like rail guns, space-battle projectiles would be fired at much higher velocities than the World War II type and, combined with no air resistance to slow them down, would reach the target far faster. Furthermore, futuristic projectiles could have built-in guidance systems that would enable them to track and follow their targets, and the guns that fired them would have computer-controlled aiming systems. World War II gunners could not simply aim at their targets and fire. They had to estimate the altitude, speed, distance, and direction of travel of their target as well as the cannon shell's trajectory and time of travel, then aim ahead of the target in the hope that the projectile and aircraft would arrive at the same place simultaneously.

Exploding cannon shells helped relieve the need to actually hit the aircraft. By exploding nearby the cannon shell could disable an aircraft with shrapnel or, if close enough, the blast's pressure wave. To do so, the time of arrival near the target had to be correctly estimated so that the exploding shell's fuze could be properly set, still a daunting task.

The fuse-setting problem eventually was solved by Americans with the invention of a proximity fuse. This device used a miniaturized radar system that caused the cannon shell to detonate when it was about 75 feet away from its target and reduced the number of five-inch shells required to down a kamikaze aircraft roughly by a factor of 3. The proximity fuse was first implemented in 1943 and was considered the third most profound invention of World War II behind only radar and the nuclear bomb.[3]

Since the transistor did not exist yet, the proximity fuse was based on radio-tube technology. These had glass housings with filaments inside somewhat similar in appearance to lightbulbs. Given the far more advanced technology that would exist in the future, the performance of cannon shells is going to be orders of magnitude better than the World War II type. These future "smart projectiles" would not just be able to detect the proximity of their target, but actually track the targets and adjust the projectile's trajectories, at least somewhat, by using onboard microcomputers and mini-thrusters.

At today's rail gun speeds, the projectile traveling at over 1.5 miles per second would give any attacking spacecraft within visual range no more than a few seconds to detect the projectile and attempt to take evasive maneuvers.[4] Given rapid-fire, multiple gun emplacements, and ultrahigh-velocity smart projectiles with computer-controlled aiming, how could defenders miss? Why would the defenders need fighter-type spacecraft of their own?

If the defenders were using laser weapons with beams traveling at 186,000 miles per second, the situation could be even worse for the attackers. What's more, since there is neither air nor significant amounts of dust in outer space, the laser beams would not show as visible lines. Attacking forces might not know they were being shot until they were hit. While the damage from using a laser to burn a hole in a spacecraft would not be as dramatic as a direct hit from a high-velocity rail gun projectile, hitting a spacecraft with multiple laser beam blasts could be devastating—much like multiple machine-gun bullets hitting a World War II aircraft. If nothing else. such blasts could damage a spacecraft's sensors or cause seriously annoying air leaks in a pressurized spacecraft.

Unfortunately, laser beams tend to spread out as they travel through space, which limits their effective range significantly. Of course, the effective range depends on the design and power output of the laser but is probably limited to a few hundred miles. Lasers also generate a significant amount of internal heat, which somehow has to be dissipated. In outer space this is a problem, since the only ways to carry heat away from a spacecraft are through radiation heat transfer or by ejecting overheated mass such as cooling water, both of which can help mark the location of the spacecraft.

Regrettably, for defenders, blowing up an attacking space ship at close range could itself be a disaster. In World War II, after shooting down attacking aircraft, ships tended to be protected from harm by an invisible force field. The force field of gravity, in combination with air resistance, would slow the forward speed of the shot-up aircraft and pull it downward into the ocean, often preventing the flaming wreck from slamming into and damaging the ship. In outer space, there would be no gravity to deflect nor ocean to absorb shot-up wreckage. Thanks to conservation of momentum, if an incoming spaceship were blown up, the pieces would still continue on their forward path, eventually slamming into the defending ship like a shotgun blast instead of a bullet. Without air resistance, none of the wreckage pieces would slow down until they impacted something.

While the convoy used by the Japanese at Guadalcanal was ineffective, it was the best option under the circumstances. Indeed, four of the transports did reach their destination and unloaded some of their cargo before being destroyed. With the right weather conditions, mistakes by the Americans, and/or dumb luck, the convoy might have succeeded. However, in outer space, traveling in a convoy formation could be an invitation to disaster. Without gravity or air resistance, if one of the spacecraft in the formation blew up it would shower the others with high-velocity debris—which would not be slowed by air resistance and conveniently fall into the ocean as in World War II.

In the science fiction universe, spacecraft would be protected by generating their own force-field shields. These supposedly would keep the pieces of exploding spacecraft from penetrating a defender's hull. Unfortunately, momentum still would be conserved during the collision of wreckage with the shields. Several tons of wreckage slamming into the shield would seriously jostle the defending ship with an effect similar to a fist slamming into the jaw of a fighter during an MMA match. While the fist would not penetrate the fighter's skull, the sudden transfer of momentum to it could cause major disruption inside. Then there is also the embarrassing fact that no one currently has a clue about how to actually build such a force-field system for shielding spacecraft. Hence, the only way to safely blow up an incoming spacecraft would be to do so at a considerable distance.

The Difficulty of Intercepting Invaders in Deep Space

Unlike the Japanese transports in the Guadalcanal battle, space ship transports would not be slow moving. To travel millions of miles from Earth to Mars would require very high speeds—for the sake of illustration, say around 20,000 mph. To intercept such a transport, an attacking spacecraft, if it were traveling in a straight line at a similar speed in roughly the opposite direction, would give a closing speed on the order of 40,000 mph. If the battle had to be conducted within visual range, the two spacecraft would have maybe two to three seconds to aim and fire as they zipped past each other and out of range.

Most likely neither the transport or the attacking spacecraft would be firing their main thrusters as they passed each other, a fact that would make them harder to detect from a distance. Unlike aircraft, spacecraft do not need to keep using their main thrusters after reaching their desired velocity. Both transport and attacker probably would have reached cruising velocity and shut down their thrusters long before they passed each other.

After zipping past the transport, the attacking spacecraft would have to turn around and accelerate in the direction of the transport if it were to continue the attack like a World War II aircraft. To do so, the attacking spacecraft would first need to use its small control thrusters to rotate itself so that its main thruster was perpendicular to the direction of its velocity, as shown in figure 13.1. It then would need to fire its main thrusters while continuing to rotate the spacecraft with its smaller control thrusters so that the force produced by the main thrusters always pointed at the center of rotation. After traveling in a semicircle, the attacking spacecraft would need to turn off its main thruster. The spacecraft would still be traveling at a speed of 20,000 mph but in exactly the opposite direction from which it came—so far, so good.

Unfortunately, with a limitation of eight g of acceleration, the attacking spacecraft would require a turning radius of over 600 miles, which would

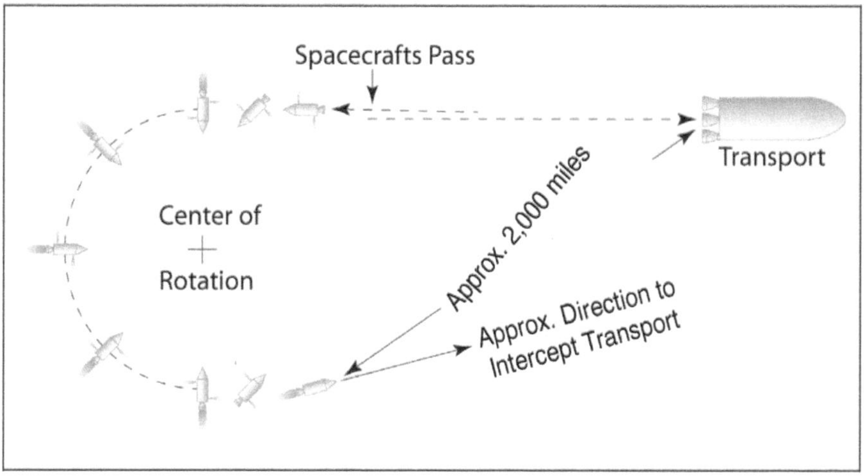

Figure 13.1. Diagram of an attacking spacecraft turning around similar to the way a World War II aircraft would turn around, assuming the attacking spacecraft and transport were both going 20,000 mph (32,187 km/hr) in opposite directions when they passed each other (not to scale). *Source*: T. K. Rogers.

leave it 1,200 miles to the side of where it had started to make its U-turn and subject both the spacecraft and its pilot to nearly six brutal minutes of high acceleration. This activity would need to be done under computer control since the pilot could very well be unconscious before the six minutes were up. Since the transport would have been moving away while its attacker was turning around, the attacker would now find itself around 2,000 miles behind its prey and need to travel faster than 20,000 mph to catch up.

However, an attacking spacecraft is not like a World War II airplane. For a spacecraft the circular turnaround is suboptimal. Instead, it would perform a maneuver that would be impossible for a World War II aircraft. When the attacking spacecraft zipped past its target, the attacking ship would rotate itself so that its main thruster could be fired to slow itself down and bring it to a complete stop, similar to slamming on the brakes in a car. By keeping the thrusters firing, the spacecraft could accelerate back to over 20,000 mph in the opposite direction, allowing it to catch up with the transport it was attacking. A little before it caught up, the attacking spacecraft would need to again rotate its thrusters to slow down if it wanted to keep from once again overshooting its target. All of this would be done with the attacker traveling in a linear rather than a circular manner.

If the attacker timed the maneuver right so that it began slowing down when the distance between the attacker and transport was about 1,500 miles, the attacker could be going the same speed in the same direction as the transport about the time the transport pulled up alongside. A maneuver of this type would have the undesired effect of clearly marking the attacker's location with the fiery discharge of its rocket motor way before it closed with the transport. It might as well train a searchlight on itself—not a good thing if someone has a reason to be shooting at you in the blackness of outer space. The good news: The stop and reverse maneuver would only require about four minutes of eight g acceleration rather than the six minutes needed for a circular turn-around maneuver; hence, the pilot would be less likely to pass out.

Given a closing speed of 40,000 miles per hour, a 50-caliber machine-gun bullet would have a kinetic energy equal to exploding four pounds of TNT when it struck its target, because the 40,000 mph velocity would be added to the bullet's normal muzzle velocity. In this situation, the machine-gun bullet would be the equivalent of a cannon shell. If the attacking ship reversed course, pulled alongside the transport, and matched its speed, a 50-caliber bullet fired by the attacker would have about 0.2 percent of the kinetic energy it would have had in the previous example with the high closing speed. Reversing course is going to make an attacker's projectile-type weapons far less deadly.

Regardless of how it was done, maneuvering to match the direction and speed of the transport, by itself, would consume far more fuel than required to blast off from Mars and make a simple one-pass attack. A human-piloted spacecraft capable of turning around or stopping and reversing velocity in outer space using its own power would be gargantuan-sized compared to a typical fighter jet on Earth.

Problems with Turnarounds

Unfortunately, if a Martian spacecraft did not have the fuel required for a turnaround type maneuver and made a one-pass attack, bringing the craft back to Mars would itself be a serious problem. A similar situation was portrayed in the 1995 movie *Apollo 13*, which described the real-life near-disaster of a 1970 Moon mission. On the way to the Moon, an explosion in an oxygen tank seriously crippled the spacecraft, leaving it barely enough electrical generation capability to run the equipment needed for returning to Earth. Since lives were at stake, time was of the essence. On this basis, immediately turning the craft around and heading it back to Earth was the desired option. But the craft was traveling at several thousand miles per hour and did not have the substantial amount of fuel needed to turn around. Instead, the spacecraft spent the additional time required to slingshot around the Moon in order to make a U-turn and head back to Earth. If a spacecraft does not have enough fuel to turn around on its own power, it needs to use the slingshot maneuver around a planet or moon to reverse its course.

In a slingshot maneuver, a spacecraft passes close enough to a planet or moon's surface so that it is captured by the celestial body's gravity and partially orbits it. To do so, the spacecraft must be going the right speed for its distance from the surface: too slow, the spacecraft crashes; too fast, it turns part of the way around, then heads off into space again. If the proper conditions have been met, the spacecraft has to fire its main thruster and speed up as it completes its U-turn in order to break free from continuing to orbit. The gravity of the planet or moon provides the centripetal force needed to make the turn, rather than providing it by burning huge amounts of fuel with the main thruster. Ironically, the slingshot maneuver is essentially the same as the circular turnaround maneuver earlier described as suboptimal except for the way the centripetal force is provided.

Unfortunately, if the spacecraft sent from Mars is traveling away at a considerable distance from its home planet when it attacks the transport, the nearest planet available for a turnaround will be Earth. The human pilot of the attacking spacecraft had better either have a big fuel tank or have packed a really big lunch.

The Penalty of Pilots

If the described one-pass attack were to occur, a much better tactic would be to replace the human pilots with computers using AI and state-of-the-art sensors then crash the attacking spacecraft head-on into the transports. However, this would be like duck hunting with a rifle. An even better tactic would be for the drone attack ship to deploy an array of javelin-shaped projectiles forming an enormous shotgun-like pattern aimed toward the transport. At closing speeds on the order of 40,000 mph, one of these javelins would cut through a transport like a plasma torch.

Another tactic might be to arm the drone spacecraft with a nice-sized hydrogen bomb. But nuclear bombs are not as effective in outer space as on Earth. Since there is no air, space-detonated nuclear bombs will not generate a massive blast wave. The lack of air also will moderate the thermal radiation. On Earth, the air near the bomb blast absorbs some of the X-rays generated by the blast, which causes the air to superheat. A nuclear explosion in outer space would mainly generate a large quantity of X-rays with some gamma rays. Detonated within at most a few miles of a spacecraft (the exact distance depending on the yield of the bomb and level of spacecraft radiation shielding), this radiation could kill the human occupants. It also might ionize some of the air or other materials inside the spacecraft enough to damage the electronics and computers, but these are easier to harden than the people. Given the expense, complexity, and limited effectiveness of nuclear bombs in outer space, they probably would not be used.

Computer-controlled (using AI) spacecraft could be designed for handling greater accelerations, making them faster and more maneuverable than a human-controlled craft. Since computers don't require oxygen, pressurized suits, or any of the other plethora of devices and gizmos needed to keep a pilot informed, in control, and alive, an unmanned spacecraft could be not just faster and more maneuverable but also much smaller, cheaper, and essentially much more effective than a manned craft, all without placing any human operators at risk.

Even NASA has been guilty of perpetuating the myth that humans are needed to pilot spacecraft. When returning to Earth, the Space Shuttle had to be under computer control as it entered the atmosphere. A human pilot could not react to conditions and make adjustments fast enough. Once it was approaching the landing strip, the human pilot typically would be allowed to bring the shuttle in for the final part of the landing, but this could have also been done under computer control. Furthermore, when the human pilot took over and used his cockpit controls, they sent electronic signals to the computer that then made the actual adjustments needed for the shuttle to land. Had the computer systems failed, the pilot would have been helpless, and the shuttle would have crashed. By contrast, had the human pilot passed out, computers still could have landed the shuttle safely.

Keep in mind that the shuttle's 1970s-era computer was significantly less powerful than one of today's smartphones. One can only wonder what a future computer with AI could do.

There are patterns that do repeat throughout history, but the World War II aircraft carrier battle model is not one of them. Even in today's Earth-bound navy, the World War II battle model is essentially obsolete. Shipboard anti-aircraft missiles and radar-aimed, computer-controlled rapid-fire guns have pretty much converted a World War II–style aircraft attack into a form of suicide. Nevertheless, there is still a pattern of history involved even here. High-energy weapon-delivery systems—from trebuchets to cannons, from aircraft-delivered bombs to guided missiles and beyond—have historically evolved toward greater range and superior targeting accuracy with less and less need for direct human input.

Where an Outer Space Fight Would Occur

The patterns of history, at least the ones associated with military conflicts, that are most likely to reoccur are those with underlying mathematical and/or physics principles. To understand the nature of underlying principles, let's return again to World War II in the Pacific. If a Japanese naval task force were going to achieve complete surprise when attacking an American-held island, it would need to avoid being detected by reconnaissance aircraft. (For the sake of argument, we will ignore detection by submarines and radar.) The probability of avoiding detection is related to the maximum size of the area reconnaissance aircraft must search to find the task force. Measure the distance from the task force to the island, draw a circle around the island using the measured value as the radius, and the area inside the circle represents the maximum area that would need to be searched to find the task force. If the task force were slightly outside the extreme range of patrol craft (100 percent chance of avoiding detection), then at half the original distance from the island, the maximum search area would be reduced by 75 percent. At one quarter the original distance from the island, the maximum search area would be reduced by 94.75 percent.

Of course, there are more factors involved in detecting an approaching naval task force than just the maximum search area—dumb luck definitely helps—but clearly, the probability of avoiding detection decreases rapidly as the task force approaches the island. And clearly, there are underlying mathematical and physics principles involved in creating the decrease.

If the same type analysis were made for the invasion of a planet, instead of searching a two-dimensional area around an island as was the case with

World War II aircraft, a reconnaissance spacecraft would have to search a three-dimensional volume of space around the planet. If the invading force were just outside the extreme limit of detection, then at half this distance, the maximum search volume of space would be reduced by 87.5 percent. At one-quarter the original distance, the maximum search volume of space would be reduced by 98.4 percent, results that are even more dramatic than the results of the previously mentioned World War II analysis.

For all that, incoming spacecraft would likely not need to be detected using reconnaissance spacecraft. The primary means of detection could well be telescopes. To quantify this, we would have to consider the number of telescopes, their resolution, magnification, depth of field, and so forth, which are beyond the scope of this analysis. Nevertheless, based on the probability of detection alone, outer space battles between invaders and defenders, if they occurred, would be increasingly likely as the invaders approached their targeted planet.

Likewise, the ability to defend a planet from invasion would be far easier near it than at great distances. The planet's occupants could readily set up an effective system of defensive satellites orbiting no more than a few thousand miles above its surface. At such distances, time lags in the transition of messages between the satellites and ground would be minimal. Even so, the devices could be set up with enough AI to act autonomously if needed. The satellites could be armed with missiles, lasers, or simply be configured to crash into invading spacecraft kamikaze-style. Thousands of grapefruit-sized or even smaller units could be sent aloft for very little money. These could be used both as weapon and sensors systems. Unlike World War II aircraft, unmanned satellites would require no fuel to remain in orbit above their planet ready to sense and destroy invaders.

Ground-based weapons could be very high-powered lasers. They could shoot at invading spacecraft with speed-of-light velocities. Since they would not need to be launched into space, the size of the lasers and the power plants used to energize them would not be issues. Ground-based missiles also could be used but would have some limitations. They would need fairly large rocket motors to get their payloads into outer space and take fairly long periods of time to reach their targets, even if the targets were in low planetary orbits. The missiles also would be easily visible during launch, which could make them vulnerable to countermeasures. By comparison, manned fighter-type spacecraft would require even larger boosters to launch them from the planet and, given human limits on acceleration, would take substantially longer to reach their targets. Sending up mass formations of such manned spacecraft would be pointless.

NOTES

1. James D. Hornfischer, *Neptune's Inferno: The U.S. Navy at Guadalcanal* (New York: Bantam, 2011).

2. Ed Jennings, "Crosley's Secret War Effort-The Proximity Fuze," accessed June 5, 2019, http://www.navweaps.com/index_tech/tech-075.php.

3. Ibid.

4. Navweaps.com, "Electromagnetic Rail Gun Proposal," accessed June 5, 2019, http://www.navweaps.com/Weapons/WNUS_Rail_Gun.php.

Chapter Fourteen

Stealth: The Key to Survival?

As explained in the last chapter, if an incoming threat can be accurately pinpointed, it can be destroyed, and space weapons are unlikely to miss. Furthermore, since there is no air resistance or gravity pulling projectiles into the ground, space weapons are inherently agile (given onboard AI guidance systems), high-velocity, long-range weapons that would be difficult to outrun or outmaneuver. Armor is one possible survival strategy, but given the high velocities of space weapons and the need to keep the mass of spacecraft reasonably low, armor's usefulness is limited. Well-thought-out damage control designs and strategies certainly would help but are only stopgap measures. The best strategy is stealth. If the location of a threat cannot be determined, the threat cannot be destroyed.

Simply put, distance equals stealth. If a spacecraft is far enough away, essentially it will be impossible to detect. For example, at a distance of 240,000 miles (386,000 km), roughly the distance to the moon, the Hubble telescope can detect, at best, an object about the size of a football field (about 93 meters).[1] To get a crude image (as opposed to a single dot), the object would need to be the size of several football fields. Move the football-sized object beyond the above distance, and the odds of detection with the Hubble approach zero. Of course, one could get a bigger and better telescope, but keep in mind the Hubble weighs 24,000 pounds (10,900 kg) and cost $2.5 billion. Using a Hubble-type telescope as a spacecraft's long-range eye is going to be problematic. There is also the fact that the amount of sky viewed declines with the telescope's magnification. To see an object 240,000 miles away with a Hubble-type unit, one has to know roughly where to look or else spend an interminable amount of time searching.

Detection also depends on the object's brightness relative to its background. Dim objects are harder to detect. Reduce the relative brightness, and the object will be able to move closer and still not be seen, and that is the essence of stealth strategy: Reduce the maximum distance for detection.

For a battle-type spacecraft to survive, it must detect and attack a hostile craft before being detected and attacked itself. For troop and supply transport spacecraft, the challenge of survival is even worse. The probability of being detected goes up exponentially as the distance between a spacecraft and its enemy decreases. Fulfilling its mission requires a transport to unload its cargo on its enemy's planet.

Virtually every method of detection depends on some aspect of using the electromagnetic spectrum. If an object is warmer than its surroundings, it will glow like a lightbulb in the infrared portion of the spectrum. Unless the object is nearly red hot, however, the glow is going to be dim and hard to detect. In the near vacuum of outer space there is only one form of heat transfer: radiation, which is easier to block than on Earth. This can be done by covering the outside shell of the spacecraft with multiple layers of aluminum foil or aluminum-coated plastic film (similar to shiny Mylar emergency blankets). Only the inner layers need to be shiny. The outermost layer could have an entirely different surface treatment. The layers would mostly not be touching each other, and the space between them would be under vacuum conditions. Try this on Earth, and air pressure outside the insulation blanket will tend to collapse the space between the layers as soon as a vacuum is created in it.

Of course, any spacecraft firing its main thruster is essentially spotlighting itself. And its plume of exhaust gases is going to be glowing in the infrared spectrum for some time after they're exhausted. Fortunately, unlike jet aircraft flying in Earth's atmosphere, a spacecraft does not need to continually fire its main thrusters to travel forward.

Almost any electronic device aboard a spacecraft, including computers, will give off some form of electromagnetic radiation that potentially could be detected. Certainly, Wi-Fi and Bluetooth devices are deliberately designed to do so. However, these emissions can be largely shielded from broadcasting any significant distance both at their source and by the outer hull of the spacecraft. Besides, they are weak signals.

Barring broadcasting and glowing, a spacecraft could only be detected by reflecting a source of electromagnetic radiation shining on it or essentially by creating a shadow when passing in front of sources of electromagnetic radiation such as stars.

CAMOUFLAGE

While *Star Trek* movies would have us believe that a secret, superhigh-tech cloaking system is required to make a spacecraft disappear in the visible light part of the EM spectrum, it's really fairly simple: Paint it black (see figure 14.1). Given the long-range effectiveness of space weapons, outer-space battles are going to be fought with the combatants separated by hundreds of miles, possibly more. At such distances, even with high-powered binoculars, a spacecraft is going to look like a dot. Put a black dot on a black background speckled with stars, and it's going to disappear. The dot could be detected as it passes in front of stars, but if it is moving several thousand miles an hour, the momentary blocking of a star's light is going to happen so fast that it will be hard to detect, even with specialized equipment.

The sensitive equipment required to see a black spacecraft (think black dot) at a large distance likely would require looking at it for some period of time using a telescope with highly sensitive detectors under computer control. But the more sensitive the optical system is for detecting dim objects, the more vulnerable it would be to countermeasures such as shining a laser into its electronic eye.

The downside of black paint is that, if aimed toward the Sun, it will absorb solar radiation and heat up. This would cause the spacecraft to glow in the infrared part of the EM spectrum, but it would be a faint glow. The glow probably could not be totally eliminated over the entire surface of the spaceship, but a cooling system could be designed to minimize it on surfaces pointed toward a planet.

The next step toward invisibility would be active camouflage. Here the surface providing concealment could change its colors, patterns, or darkness

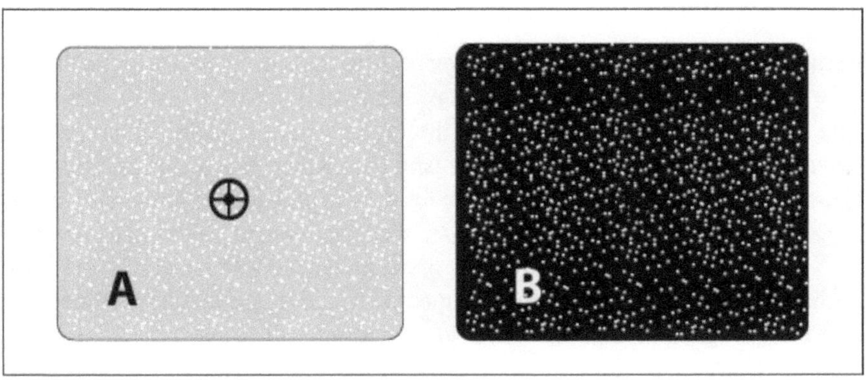

Figure 14.1: A) Black spacecraft on shown on a grey background. B) The same black spacecraft shown on a black background. *Source*: T. K. Rogers.

to match its background. An image of what's behind the object would not be passing through it, and so it would not be possible to do something like read the words on a sign (if one were there) behind the camouflaged item. For natural backgrounds like rocks, clouds, and stars, active camouflage allows the concealed object to get even closer before being detected.

Getting closer was one of the goals of American antisubmarine aircraft during World War II. German U-boats moved slower underwater than on the surface and could only remain submerged for relatively short periods of time. Hence, they usually operated on the surface. When an aircraft attacked, the U-boat could survive if it submerged before the aircraft arrived, otherwise the U-boat would be in serious danger. The Americans sought to address this by developing the Yehudi lights.[2] These were forward-facing lights attached to the front fuselage and leading wing edges of antisubmarine aircraft. The lights used photocells to automatically adjust their output so that it matched ambient light levels, essentially a form of active camouflage. Based on trials, the system typically reduced detection of the aircraft from about 10 miles down to less than 2 miles. Even though they worked, Yehudi lights were not deployed. They were made obsolete by advances in radar.

In its most developed form, active camouflage would be like putting a video camera on the back side of a spacecraft and a gigantic screen TV on the front. For best effect, the sensors of the video camera likely would be distributed over the surface of the spacecraft and be connected to corresponding pixels distributed on the other side. At a distance, this could reduce the spacecraft's detectability in the visual spectrum but not reduce it in the entire EM spectrum. At some point, the glowing pixels might heat up enough to be detected by an infrared scan.

The Pros and Cons of Radar

Radar is similar to turning on a powerful searchlight in a dark night. In this case the searchlight is emitting microwave radiation instead of visible light. The microwaves are reflected from the target back to sensors that can then detect the target. One of the key advantages: Radar has a wider beam that is not affected by clouds, smoke, or dust, and thus can illuminate objects at much greater distances over a wider area than a searchlight can. Another advantage (although one we humans are reluctant to admit): Radar systems, unlike eyes and the human brain connected to them, are machines, ones that don't get fatigued or have moments of inattention. Since it's a machine, radar also can characterize incoming threats in terms of numbers—for example, an incoming aircraft at 10,000 feet altitude, closing at 270 mph on a heading of

82.7 degrees, at a distance of 12.4 miles. And numbers pinpoint positions, the first step in destroying targets.

A key disadvantage of using radar: Like turning on a searchlight, radar can announce one's position. During the Vietnam War, U.S. aircraft used air-to-surface missiles designed to track radar signals coming from anti-aircraft installations, follow them to their source, and blow up in close proximity.[3] Of course, the people at the antiaircraft installations soon learned to turn their radar off and on in ways that confused the incoming missiles. For every attack, there is a countermeasure and for every countermeasure a counter-countermeasure . . . ad infinitum until someone gives up (or dies).

Radar can also be defeated by jamming—in other words, by overwhelming the receiver that is supposed to detect reflected radar beams. Jamming is like shining a bright light into the eyes of the gunners attempting to use a searchlight to see an aircraft on a dark night. The bright light would blind them to the dim reflection coming from the aircraft. Once again, the sources of jamming signals can be tracked and potentially destroyed, which means there will be yet another chain of countermeasures dealing with countermeasures.

Spacecraft might use forward-looking radar for spotting and avoiding space junk. In outer space where there are no clouds to obscure vision, a spacecraft on a military mission could turn off its radar with relatively little risk of having a collision with random space junk or a failure to detect an enemy. First, there isn't a lot of space junk floating around; and second, most of it, especially the big objects, could be detected about as well with automated vision systems as with radar. Certainly, an enemy spacecraft would qualify as a big object.

A spacecraft could avoid detection from radar by painting its surfaces with radar-absorbing materials—similar to using black paint to absorb visible light—and designing spacecraft surfaces so that they reflect radar beams away from the sensors designed to receive the reflected signal. Current technology already has made warfighting aircraft nearly invisible to radar, but such aircraft have to be flown on moonless nights to avoid visual detection. Ironically, this has resurrected interest in Yehudi light systems. Certainly, spacecraft could employ multiple technologies and systems to avoid detection in the various parts of the EM spectrum. Since there's no air in outer space, the shape or surface texture of a spacecraft would not be limited by aerodynamic considerations. Of course, as mentioned before, all a stealth technique can do is let an attacker get closer to its target. Eventually, the attacker will be detected, and the fight will be on.

Confusion, Cover, Concealment, and Deception

When the bullets begin to fly, what do real men do (at least in war movies about infantry engagements)? They shout, "Take cover"—as if it needed to be said! Oddly enough, it took most of the Civil War for its generals to understand that it was not a good idea to walk in formation toward an enemy that had taken cover behind barriers like walls and were armed with what for their day were accurate long-range weapons. The error was understandable given that in the previous conflict, the Mexican War, the smoothbore muskets of the time were ineffective beyond a range of 100 yards as compared to a maximum range of around 500 yards for muzzle-loading Civil War rifles.[4]

Cover, like armor, reduces a target's vulnerability to weapon fire by absorbing damage that otherwise would be done to the target. On the other hand, concealment makes the target harder to see and accurately pinpoint but offers no actual protection from weapon fire. Ducking behind a wall next to a doorway in a typical house when engaged in a gunfight is an example of concealment but not cover. Even handgun bullets will easily go through the average wall in a house. Still, a wall does introduce uncertainty into the act of aiming for a killshot.

Ducking behind a car during a gunfight can be either cover or concealment depending on one's location: Behind the engine equals cover; behind doors equals concealment. Unfortunately, even handgun bullets can go through car doors. As usual, Hollywood has mucked it up with things like chase scenes in which a jeep mounted with a 30-caliber machine gun chases the hero's unarmored car while peppering it with dozens of bullets—each substantially more powerful than a typical handgun bullet. In real life, these would go through the trunk, the seats, and the people in them. Some of the bullets might end up wrecking the car's engine.

In World War II naval battles, when cannon shells started flying, taking cover was typically not an option; however, concealment was, at least to some extent. The Battle off Samar—part of the Battle of Leyte Gulf—which occurred when U.S. forces began the Philippines invasion, provides a good example.[5] The main American naval forces had been lured away leaving only a small force, Taffy 3: three destroyers, four destroyer escorts, and six small-sized escort carriers (13 ships total). About that time a major Japanese navy battle group inconveniently showed up with four battleships, six heavy cruisers, two light cruisers, and 11 destroyers (23 ships total). A single ship, the *Yamato* (the largest battleship in history) had a displacement roughly equal to that of the entire Taffy 3 force. The *Yamato* bristled with guns, including nine 18-inch cannon as compared to Taffy 3, whose biggest guns were five-inch cannon that were not capable of penetrating the hulls of the larger Japanese ships. The small aircraft carriers were ordered to launch their aircraft, then head for a nearby squall in the hope that it could conceal them. Meanwhile, the other American ships laid down a smoke screen to help hide the carriers' retreat.

In the ensuing battle, instead of sensibly retreating, the American destroyers, destroyer escorts, and aircraft attacked. The plucky Americans did manage to sink or disable three heavy Japanese cruisers, but for the most part simply harassed the Japanese battle group. It was like Chihuahuas attacking Great Danes. Yet by the end of the battle, in a stunning reversal of fortune, the Great Danes retreated.

Why the Japanese withdrew is a mystery. Certainly, the battle was not won with smoke screens, but they probably helped. The Japanese did have radar to use for fire control, but their best form of fire control depended on optical equipment. Possibly the smoke screens combined with the vigorous American ship and aircraft attacks interfered with the Japanese force's ability to aim and to evaluate the strength of the opponent. The excess fuel they consumed maneuvering during the battle and the uncertainty of resupply may have played a role in the Japanese withdrawal. Nevertheless, the American tactics illustrate how a vastly superior force with long supply lines can be defeated with a greatly inferior force.

For the World War II equivalent of hiding ships behind smoke screens, spacecraft could fire out vast quantities of balloons. In an airless environment, these would require very little gas pressure to inflate and could be fired at incredibly high velocities. Equipped with low-cost microchips and small thrusters, the balloons could flock together and form vast maneuverable target-obscuring clouds or act as decoys for drawing fire away from the real spacecraft.

Fill the balloons with a fast-expanding foam and they would not pop even if hit by laser beams. The balloons could be designed to reflect, block, or absorb various parts of the EM spectrum, thereby disrupting an enemy's ability to detect targets. Since they could be fired by the hundreds at high velocities toward hostile spacecraft, the balloons might look like incoming projectiles, a tactic that could confuse enemy countermeasures and cause them to maneuver out of their desired firing positions.

Inflatable balloons have been used as decoys in nuclear warheads atop ICBMs as a means of confusing enemy radar and making it nearly impossible to detect and destroy incoming warheads. The time window for destroying incoming ordnance is generally small and the available equipment for doing so, limited. Numerous decoys resembling incoming projectiles can overwhelm countermeasures by giving them too many possible targets other than the real one. Likewise, decoys resembling hostile spacecraft could overwhelm targeting systems.

Even rail-gun projectiles can be destroyed by weapons such as lasers before impact if they can be accurately detected and tracked. Due to their much smaller size compared to a spacecraft, projectiles have a degree of natural stealth. They also could benefit from absorptive coatings and especially from measures taken to keep their surfaces cold in order to prevent them from glowing in the infrared spectrum. This is one of the problems with using rocket-propelled projectiles. Not only would they have a significant heat signature making them easy to detect, but they also would leave a trail of burned fuel pointing back at the location of the spacecraft they were fired from. The best way to use rockets in outer space would be to fire them from drone spacecraft. That way, the main spacecraft's location could remain obscure.

Rockets do have at least one important advantage: They don't produce recoil in the spacecraft launching them. On the other hand, rail guns would have a significant recoil that could knock a spacecraft firing them off course. They mostly would be appropriate as weapons onboard larger spacecraft. For a given projectile's mass and velocity, a rail gun will have less recoil than a cannon because the rail gun will not expel the considerable mass of burning gunpowder that a cannon will. However, there is no way to fully eliminate a rail-gun's recoil.

The photon torpedoes used in the *Star Trek* franchise are an example of a silly, easily seen, somewhat slowly moving projectile. Why would a crew in a well-armed, highly advanced spacecraft sit around hoping their shields would hold while a glowing orb drifted toward their spacecraft? How could they not target and destroy it well before it arrived?

Still, it may not be necessary to pinpoint the exact position of incoming projectiles to deflect or destroy them. Decoys like the inflatable balloons mentioned above could lure projectiles with tracking capability away from their intended targets, but they also might be capable of disabling incoming projectiles. These could be used against thinly armored spacecraft. If the projectiles penetrate too easily, they simply would travel through the targeted spacecraft while leaving a relatively small entry and exit hole. Certainly, this would be harmful, but with damage control measures, would be unlikely to totally disable a spacecraft. To be effective the projectile would need to break apart and quickly release its kinetic energy or detonate any explosive charge it carried almost the instant it penetrated the spacecraft's wall.

Fired at rail-gun velocities, a 2.1-gram balloon (the same mass as a .22 long rifle bullet—pretty much the smallest hunting rifle bullet made) would have about the same kinetic energy as a 50-caliber machine-gun bullet. While the balloon would have little penetrating ability, its impact could set off the explosives in the incoming projectile or cause it to break up and release its kinetic energy prematurely. The balloon might damage the incoming projectile's tracking sensors, rendering them useless. Certainly, given their small amount of mass, it would be no problem to fire a shotgun-like pattern of balloons at an incoming projectile.

Timing would be the key to using shotgun-like patterns of balloons as an effective countermeasure. Fired too early or too late, the balloons would be worthless. To be effective, defenders would need to know exactly when attackers fired their weapons. Given the size of the balloons and the number fired, they would be unlikely to miss.

Could a pattern of balloons fired toward an incoming laser pulse be an effective countermeasure? The answer is a definite maybe. Lasers need to stay focused on a target for at least a short time to burn through it. The balloons could be designed to absorb, reflect, or diffuse a laser beam, thereby moderating the burn-through effect. Certainly, this could help. It would be too late to totally avoid being hit by the laser if the balloons were not fired until the laser pulse was detected. Nevertheless, it would be better to fire the pattern of balloons after taking a laser hit than to not fire and end up taking numerous laser hits or more lengthy ones.

Creating a cloud of balloons around the defending spacecraft would do a better job of providing cover against laser beam attacks than firing them at the laser, especially if there were multiple layers of balloons in the cloud. In this case it would not be necessary to detect that a laser had been fired before launching the balloons. It could be difficult to keep the balloon cloud in place for a lengthy period of time. When the balloons were launched, they would have the exact same velocity as the spacecraft launching them, along with whatever additional velocity the launcher provided. Given some flocking ability, the balloons would be able to surround and stay with the spacecraft that had launched them, as long as the spacecraft did not fire its main thrusters. If that were to happen, the balloons would not have enough thruster power to match the velocity changes created in the spacecraft.

The Harry Potter Invisibility Cloak

The invisibility cloak depicted in Harry Potter movies could be drawn around a person or object and render it completely invisible. In other words, an image of anything behind people wearing such cloaks would appear to be transmitted through them with no distortion. For military purposes, the cloak would have to hide the person's EM emissions created by body heat and need to be transparent to all forms of radar.

Thanks to the popularity of Harry Potter, the internet is full of posts declaring that any device even remotely related to active camouflage is somehow a top-secret invisibility cloak. There are the honest small-scale demonstrations of clever invisibility effects that could never be scaled up for a military application. Then there are the outright Photoshop frauds. Between hyperbole and fraud, it's hard to know if the military, or anyone else for that matter, actually has anything similar to an invisibility cloak. If the military really did have one, would it openly talk about it?

Metamaterials can bend microwaves around an object, thereby rendering it invisible in that part of the spectrum.[6] For it to work, the size of the hidden object must be similar to the wavelength of the EM radiation bent around it. Unfortunately, the wavelength of visible light is very small, meaning that most of the stuff metamaterials can hide in the visible spectrum are already too small to see with unaided human eyes. For that matter, given the current state of the art, a spacecraft would be too large to hide in the microwave portion of the spectrum. The technology is still very new, so who knows? Maybe further development may be possible.

There is also the possibility of invisibility cloaks based on quantum physics or various manipulations of the space-time continuum. Presently, these devices are just theoretical. By the time they become a reality, we may be traveling through space using warp drives. By then, things like rockets and rail guns may seem like throwing spears and rocks.

The Landing—Where Stealth Tends to Fail

Unfortunately for the Earthlings riding to war in a public transport ship, the Martians would have an idea where they were and when they would arrive. The size of the transport would limit its ability to maneuver or alter its orbit. The transport would be a robust vessel with an elaborated damage control system and armor capable of protecting passengers from the ship's high-speed collisions with small pieces of space junk. It also could be armed and accompanied by smaller, much more maneuverable warships to protect it. It could be fitted out with the ability to create balloon clouds or decoys and maybe a few rail guns or high-powered lasers. However, no matter how it is equipped, or even if it is a military rather than a public space transport, as the craft approached closer to Mars it would become increasingly vulnerable to detection and attack.

The most vulnerable moment would come when the invaders attempted to land. The landing spacecraft would significantly heat up as they entered the atmosphere, causing them to glow like lightbulbs in the infrared part of the EM spectrum. At that point, all hope of stealth would be lost. To make matters worse, the landing spacecraft likely would use some combination of rockets and aerodynamics to slow their landing velocity. Using rockets would make them highly visible. Relying on aerodynamics to glide in for a landing would require some form of prepared landing strip.

The Allied invasion of France on D-Day during World War II used gliders to land troops and equipment behind the German lines without the benefit of prepared landing strips, but also with many crashes and mishaps. The gliders had large wingspans, slow speeds, and lightweight construction that did not need to slow down from speeds of thousands of miles per hour and withstand the rigors of an atmospheric entry. They subsequently had a much better chance of surviving a landing on an unprepared strip than a spacecraft designed to complete its descent as a glider.

The gliders came in for landing silently at nighttime, making it hard for antiaircraft defenses to detect and destroy them. By contrast, incoming spacecraft would not only glow as they entered from heat generated by atmospheric friction, but they also would be traveling faster than the sound barrier and announce themselves with a nice loud sonic boom.

If the Martians mounted a defense against a landing, an Earthling force would need to bombard and destroy the defenses ahead of the landing (not easily done) or, failing that, have an overwhelming number of landing spacecraft and decoys, too many for defenders to destroy. The invaders might be able to land undetected in a remote part of the planet, but this would require them to transport their supplies, equipment, and themselves over considerable distances.

The best protection, at least for the initial wave, would be surprise based on a good deception instead of depending on stealth, armor, or armament. Indeed, nothing beats a good lie (we're from Earth's government, and we're here to help you) if attackers can pull it off just long enough to get their forces positioned on the ground and in control of landing facilities, they will be able to bring in reinforcements.

NOTES

1. Frank Summers, "Angular Resolution and What Hubble Can't See, Hubble's Universe," accessed June 5, 2019, http://hubblesite.org/explore_astronomy/hubbles_universe_unfiltered/blogs/angular-resolution-and-what-hubble-cant-see.

2. Vannevar Bush et al. "Camouflage of Sea-Search Aircraft: Visibility Studies and Some Applications in the Field of Camouflage," Office of Scientific Research and Development, National Defense Research Committee, 1946, accessed June 5, 2019, https://apps.dtic.mil/dtic/tr/fulltext/u2/221102.pdf; and Sebastien Roblin, "That Time the Allies Engineered a 'Cloaking Device' during World War II," accessed June 5, 2019, https://taskandpurpose.com/allies-cloaking-device-world-war-ii/.

3. Warfare History Network, "How the U.S. Military Went to War against Vietnam's Radar and Air Defenses," accessed June 5, 2019, https://nationalinterest.org/blog/the-buzz/how-the-us-military-went-war-against-vietnams-radar-air-25034.

4. Despite the increased accuracy of Civil War-era rifles, most combat took place at relatively short range. See Paddy Griffith, *Battle Tactics of the Civil War* (New Haven, CT: Yale University Press, 1989); and Earl J. Hess, *The Rifle Musket in Civil War Combat: Myths and Reality* (Lawrence: The University Press of Kansas, 2008).

5. James D. Hornfischer, *Last Stand of the Tin Can Sailors: The Extraordinary World War II Story of the U.S. Navy's Finest Hour* (New York: Bantam, 2004), and Shahan Russell, "The USS Johnston and its Kamikaze Captain, Took on 4 Battleships, 8 Cruisers, and 11 Destroyers at Leyte Gulf," accessed June 5, 2019, https://www.warhistoryonline.com/world-war-ii/uss-johnston-kamikazecaptain.html.

6. Ken Kingery, "Beyond Materials: From Invisibility Cloaks to Satellite Communications," accessed June 5, 2019, https://stories.duke.edu/beyond-materials-from-invisibility-cloaks-to-satellite-communications.

Chapter Fifteen

Arming the Resistance: They're Coming for Us; Break out the Printers

With no reason for a standing army or even for hunting, Martians would have little reason for firearms, but given a few months' warning of impending attack and the rebels' superior level of motivation (panic), Martians could 3-D–print weapons. The original 3-D plastic gun, the Liberator (plans released on the Internet in 2013), which created a firestorm of media-driven hysteria, was an inaccurate single-shot pistol that easily could have broken or blown up before firing a single box of ammunition—hardly a military-grade weapon.

The Martians would have more sophisticated 3-D printers capable of making metal parts. They also would have CNC machines and, honestly, even today a half-decent CNC machine can make a working copy of just about any firearm currently in existence (although making the firearm's barrel might also require a metal lathe). Manufacturing of this type needs to start with a block or bar of steel, but this would be no problem on a planet with a steel industry. Furthermore, there likely would be better fatigue-failure resistance in a CNC-machined firearm versus a 3-D–printed one. A CAD (computer-aided design) drawing of the desired firearm's parts would be needed, but some are publicly available or can be produced from old blueprints or from an actual firearm by using scanners. Hence, making 20th- or early 21st-century infantry weapons would be doable. Even so, making and assembling them fast enough might be a serious issue. Manufacturing more advanced 23rd-century military weapons may not be as doable. Most likely, blueprints, CAD drawings, or the actual firearms would not be easily acquired. These weapons also might include computer chips and other elements that CNC machines or even 3-D printers could not by themselves fabricate.

Ammunition is yet another issue. The Martian chemical industry would have the ability to make gunpowder, but 3-D printing cartridges would be another matter. If the Martians produced 10,000 firearms, they would need several million cartridges to feed them. Keep in mind that a single fully automatic weapon can fire at rates of 450 to over 1,000 cartridges per minute. Ammunition production would need some form of high-speed dedicated manufacturing lines—with some ingenuity they could be converted from lines already set up for a different product. For example, in the early days of modern Israel, the Israeli military converted lipstick factories into ammunition manufacturing facilities.

While personal firearms based on 20th- or early-21st-century weapons would still be deadly, they would also be obsolete—and not just the weapon, but also the projectile. Fully up-to-date ammunition could well be higher tech than the firearms that shoot it, especially ammunition for long-range sniping. Bullets of this type would contain miniaturized sensors and guidance systems allowing them to track targets and significantly increase strike probability.[1]

RESISTANCE IS FUTILE—WELL, MAYBE NOT

Tactics, generalship, and even dumb luck are all important, but few things win battles with greater certainty than superior military technology combined with its handmaiden: timely supply. Without timely supply, superior technology is next to useless. The German World War II ME 262 jet fighter is a good example. It could fly faster and was more heavily armed than its competition. By most measures, it was a game changer and yet, it made no significant difference in the air war over Europe. The jet came too late in the conflict and could not be produced in enough quantity.

On the other hand, the U.S. Navy's Hellcat fighter showed up in the World War II Pacific in plenty of time to make a difference and in enough numbers to help sweep the Japanese navy's less capable Zero fighter from the sky. The Hellcats were not just faster but also equipped with armor and self-sealing gas tanks, things the Japanese aircraft lacked. Subsequently, Hellcats shot down 13 Zeros for every Hellcat the Zeros shot down.[2]

Starting with philosophy, many factors combine to produce superior military equipment. The philosophy of design for the Zero placed little value on protecting the pilot; hence, there was no armor plating. This made the aircraft lighter and more fuel efficient, which favored maneuverability and range. Presumably a pilot would not need to be protected by armor if he could outfly his enemy and attack from distances his enemy couldn't match. Theoretically, a pilot would fight more aggressively if he knew he would be killed if he didn't kill first. But even dying had a good point: To die for the emperor was supposedly a high honor.

By contrast, the American Hellcat was virtually a flying tank designed around a philosophy based on a combination of pilot survival and brute force. If the aircraft needed to be heavier to protect the pilot and needed to fly faster to escape being shot down, then give it the biggest, baddest engine ever. So what if it drank a lot of fuel? This was the American way. The idea of an American dying for his president was laughable. For that matter, the idea of an American dying in combat for anything was aberrant.

The quality of information for engineering everything from design to manufacturing and the quality of information about enemy technology also are key factors in producing superior military equipment. These factors are magnified by using resources of time and money for military R&D. In this regard, a well-funded and focused development advantage of about a decade imparts an almost insurmountable advantage in technological superiority. The Battle of 73 Easting, fought between the American-led coalition and Iraqi armor forces on February 26, 1991, provides an example. It pitted various armored vehicles against each other, primarily the U.S. Abrams tank (fielded in 1980) against the Soviet T72 tank (fielded slightly less than a decade earlier in 1971). The American tanks were superior in almost every way. They had excellent sighting systems, could shoot accurately on the move, and could shoot at longer range than the Iraqi tanks. The Americans also combined excellent training, tactics, and leadership with their outstanding equipment, but even these things were influenced by R&D. At the end of the day, Iraq had lost 160 tanks and 180 armored personnel carriers along with a number of other vehicles and artillery pieces compared to the American loss of a single Bradley fighting vehicle.[3]

With no budget for military equipment, R&D, or personnel (other than the cyber corps), things would look grim for the Martians, but the invasion of Mars and subsequent war with Earth will not look like 20th- and early 21st-century conflicts. First, the line of supply for the Earthling invaders is going to be the longest in history. This effectively will preclude the use of tanks, heavy artillery, close air support, and the usual variety of transport vehicles. Invaders would mostly be limited to small arms. At that, they would need to conserve their ammunition since resupply time could take months.

The Earthlings will be invading territory they claim to have financed and have now come to repossess from its delinquent tenants. Not only would the invaders not want to damage and subsequently devalue their investments, but they also likely would need to use at least some of the resources generated by their investments for their survival when on Mars. This includes food, shelter, and possibly even oxygen.

It took 168 years from the establishment of the first British colony in America to the start of the Revolutionary War, and it could take about the same length of time from the establishment of the Martian colonies to the invasion by Earthlings. So, allow time for the initial exploration of Mars before serious colonization begins, round off a little, and we can assume the Martians will rebel roughly 250 years from today. Consider that electric lighting, cars, aircraft, computers, smartphones, and so on did not even exist 250 years ago, we can assume that Martian colonies will be unimaginably high-tech 250 years in the future, even though the technology may be largely imported from Earth.

Nevertheless, Martians are not going to rebel until they are self-sufficient with respect to food, energy, materials, and any other goods essential to the continued existence of their industries and prosperity. For all that, the invasion will occur before Mars is fully terraformed. The planet will have been warmed up significantly, which will increase atmospheric pressure. The atmosphere will contain some oxygen, but CO_2 will be too high and pressure still too low to be breathable without specialized breathing apparatus. (Fully terraforming Mars could take, optimistically, over a thousand years.) The Martians will be used to this situation; the invaders will not.

This thicker atmosphere will be a mixed blessing for invaders. They will still need pressure suits, though not the bulky suits used when Mars was first explored. The average temperatures will be warmer, but the thicker atmosphere will cause higher heat loss from the invaders' bodies. They will feel the cold more. Equipment that works perfectly on Earth may not function as well in the cold, dry, dusty, lower-pressure, higher CO_2 environment of Mars. On the other hand, Martian equipment will be designed for it. A similar situation happened when winter arrived during the German invasion of Russia in World War II. Because Russians were used to their winters, they were better prepared for them with everything from warmer clothing to better lubrication for their equipment. Obviously, Germans knew it got cold in winter and got very, very cold in Russia. They even had the historical example of Napoleon's difficulties with the Russian winter. Still, they failed to grasp basic details that led to equipment failures and massive noncombat-related casualties.

While they might not have cannons, Martians will have everything required to make accurately guided missiles capable of hitting any location on their planet. Given their experience in using explosives for mining and construction projects along with their general technical expertise, Martians will be masters of making and delivering improvised explosive devices. They will have numerous spacecraft for servicing asteroid mining and rocket ships for rapid travel to distant places on Mars. If Earthlings are foolish enough to concentrate their forces any place outside of the cities, the Martians can load an existing rocket-propelled craft with extra fuel and cases of explosives, then launch it so that it crashes into the Earthling forces.

In addition, the art of long-range sniping will reach a new level of deadliness on Mars. Not only will there be smart bullets that enhance the probability of kills, but sniping will be easier to do even with conventional ammunition. With Mars's lower gravity and lower atmospheric pressure, bullets will have much flatter trajectories and lose significantly less velocity, hence arriving at their destination far faster than on Earth. True, the long-range sniping advantages offered will be available to both defenders and invaders; nevertheless, the odds favor the defenders, thanks to their superior stealth capabilities. Even with real-world versions of the fictional Harry Potter invisibility cloak, it will be easier to conceal a single sniper team than an entire army on the move. Martian snipers also will be able to fire and disappear into spider holes or tunnels before the Earthlings know what is happening.

Martians will block invaders from using Mars's planetwide GPS, down-looking satellite surveillance, and satellite phone systems while defenders will have full access. Given all these factors, Martians will be able to detect and destroy any concentration of invaders anywhere on the planet. In open combat, where there is little risk of serious collateral damage to noncombatants and to infrastructure, Martians will have a clear upper hand over Earthling invaders.

Martians will have both vehicles that can be used as troop transports and earth-moving (or should we say Mars-moving) equipment that could be outfitted with armor and weapons, similar to the way tanks were first created. Martians might have advanced models of any aircraft that might be usable there. While they might not have been designed as military aircraft, certainly they could be modified. These factors will give Martians the ability to concentrate fighting forces on hot spots where they are reluctant simply to use brute-force guided missile attacks for wiping out invaders.

As for the superiority of the invaders' military R&D, the invasion likely will be an American-led coalition, and Americans have questionable talent for keeping military R&D secrets. The most critical military R&D secret ever, the Manhattan Project, which developed the atomic bomb, is a good example. Nobel Prize–winning physicist (and renowned trickster) Richard Feynman used to go out the main gate of the super-secret atomic bomb-making facility at Los Alamos, New Mexico, and return back into it through a hole in the fence. He would repeat the process again and again until it completely befuddled the guards at the gate.[4] Thanks largely to at least a half-dozen Soviet spies imbedded in various aspect of the Manhattan Project—including the infamous Rosenbergs, who were tried, convicted, and executed for revealing atomic bomb secrets to the Russians—not to mention Soviet spies embedded in the U.S. State Department, Treasury, and Office of Strategic Studies (OSS), the Soviet Union was able to explode its first atomic bomb in 1949 without having to waste massive amounts of money and time on an R&D effort.[5]

Given Mars's cyber-spying efforts and the numerous Martian sympathizers on Earth, not to mention Earthling press releases about weapon developments, Martians will know more about Earthling weapons and tactics than Earthlings will know about Martian ones. Martians will not be able to do something as complex as create an atomic bomb in the time they have to prepare for an invasion, nor be certain enough about Earth's intentions to attempt intercepting and destroying the invaders in deep space. So the invaders will arrive without being attacked in transit. But considering Earthling preparation and travel time, Martians should have around two years to prepare for the conflict (assuming that they are not too double-minded about the warning signs), enough time for Martians to be well prepared to fight on the surface of their home planet.

The Video Game-Like Conflict

Unlike battles of the 20th and early 21st centuries, frontline troops (cannon fodder) will be composed mostly of purpose-built military robots (warbots) and H-bots. Some warbots will be designed for suicide missions. Those designed for survival will be mercilessly destroyed by Martians, not just because they are threats but because they obviously are subhuman machines. The exception might be if they can be captured and reprogrammed. Martians subconsciously will be less aggressive in destroying H-bots because of their human appearance. It turns out that most humans have an inborn aversion to kill other humans. According to S. L. A. Marshall writing in his famous 1947 book, no more than 25 percent of American infantrymen actually fired their weapons in World War II when engaged with the enemy.[6] While the reliability of Marshall's data has been questioned, the aversion to shooting and killing is real, according to Dave Grossman, and has been experienced in conflicts involving firearms well before World War II. The low firing rate and the aversion to killing can be alleviated by psychologically based training.[7] With respect to shooting, this training has evolved from using round stationary targets (World War II and before), to pop-up targets, all the way to highly sophisticated virtual-reality systems. Subsequently, firing participation in combat has increased from 25 percent to over 90 percent.

At first glance, it appears that the invading humans would be much better trained in shooting and killing than the Martians, but this may be an illusion. The list of recreational facilities unavailable to Martians is lengthy: water parks, ski resorts, sports stadiums, fishing lakes, sailboats, and so forth. To compensate, Martians probably will develop some of the best 3-D virtual-reality facilities in the solar system. These would include "shooting" scenarios that offer a degree of military training and could easily be modified to intensify it.

When it comes to AI, smart sensors, and robots, Martians will be one of the most advanced groups in the solar system. Martians will depend on robots for space mining as well as for daily support. While these will not be military robots, converting them or building new warbots will be doable. Warbots will not be giant lumbering machines as sometimes depicted in movies. They will come in a wide range of sizes, types, and capabilities. Tiny smart-sensors (smart dust) the size of grains of sand (currently about the size of a grain of rice) will have limited mobility but be capable of forming ad hoc networks for collecting and transmitting intelligence information. These could be sprinkled about to monitor enemy activities.[8] Insect-sized warbots will be designed for missions such as locating and marking targets for destruction. Rat-sized warbots—a mechanized version of AI-enhanced hand grenades—could locate targets, crawl close to them, and detonate. Even the humble land mine could be turned into a type of warbot. It could fire its explosives at a nearby target rather than waiting to be stepped on. Given some capability for mobility, it could crawl forward and plant itself in the midst of enemy camps. Certainly, warbot-type mines also could be connected in a wireless network allowing them to be turned on and off or to become sensors.

Of course, there also will be heavily armed and armored warbots designed to attack and destroy. Ones of this type in the arsenal of the invaders will need to be lightweight and relatively small to be transported. Martian warbots, on the other hand, will have no such restrictions. They nonetheless will need to be relatively simple in design and easy to manufacture, since the available time will be limited. The strategy for Martian warbots in some ways will resemble that of Russian T34 tanks in World War II—designed to be mass-produced quickly with little regard for craftmanship or battlefield superiority beyond basic needs. In many cases, the T34s destroyed superior German tanks by overwhelming them with sheer numbers.

Given the many automated weapons, Martian battles will resemble video games with humans fighting from behind computer screens. Direct human-on-human fights mostly will occur in special cases or when there has been a breakthrough facilitated by the various forms of bots, or in the extreme case, if both sides run out of functioning warbots and H-bots. The initial goal of the invaders will not be to destroy the rebels but to capture their leadership and dissuade their followers from further rebellion. These are, after all, the people who will be needed to keep the Martian economy producing for their Earthling masters once the rebellion is crushed.

The desire to exploit and control Mars's resources for Earth's benefit will be a major factor leading to the invasion but probably will not be Earth's stated justification for it. From the Martian perspective, the triggering event could center on the desire to continue terraforming Mars until the planet's atmosphere is fully breathable. This means radically altering the surface of Mars by creating conditions for liquid water to exist on it, altering weather patterns, adding microbes, and massively using oxygen-producing cyanobacteria, plants, and lichens. These changes will permanently alter Mars in ways that seriously harm possible scientific discoveries about the planet's past. According to Earth's current international treaties, Mars belongs to all of humanity, which implies that Martians do not have a right to unilaterally alter their planet, or for that matter, control how its resources are exploited.

For Martians, interference with efforts at altering the atmosphere to a breathable state would be an egregious violation of the right to breathe, an emotional issue likely to incite violence. Mars will have some scientific organizations with Earthling scientists on temporary missions who virulently oppose the possibility of obscuring or destroying scientific evidence about Mars's past. They may be subjected to harassment, death threats, and violence. Mars also would contain other temporarily stationed Earthlings such as space or mining officials, maybe even some well-heeled tourists. These various Earthling visitors will provide yet another hackneyed excuse for an invasion: They will "need" protection from enraged Martians.

The Penalty for Losing

Unlike the American Revolution, the hanging offense of treason will be off the table, but captured Martian rebels still will be at the mercy of the Earthling-led invasion coalition. The coalition leaders will have full authority to detain rebels indefinitely. Due to the unbreathable Martian atmosphere, the invaders will not be able to pitch tents outdoors to form a barbed-wire enclosed detainment area. Still, given 250 or more years of future development, the invaders probably will have devices that allow them to confine prisoners to their quarters or, worse, to turn them into docile and obedient servants. In fact, it may be possible to reprogram not just H-bots but also, to some extent, humans.

On the other side of the conflict, with over two million people, Martians will have similar or possibly even better human control and reprogramming technology. Martians will have a small police force with very limited facilities for holding prisoners. While most Martians will be good citizens, some will be born with criminal or psychopathic tendencies. Some will go insane, especially considering the confining nature of life on Mars. Some will lose control of their emotions and resort to acts of violence. Some will exhibit antisocial behaviors from drug or alcohol addiction. Martians will need methods of dealing with such situations, and building a prison will not be an efficient option.

Human control devices such as collars, ankle bracelets, or implantable chips that render people unable to resist would be an excellent alternative to the massive resources needed for prisons. While this might sound like science fantasy, consider the fact that even today an ordinary smartphone can track a person's location and record audio and video while sending data to a remote location. With properly placed electrodes, smartphones could deliver a brain-scrambling or heart-stopping shock. Remove the unneeded display screen, and the device could be the size of a wristwatch, maybe smaller. Attach the control device so that it is not easily removed, and the person it is attached to could be remotely controlled, albeit in a somewhat crude manner. With RFID tags placed in doorways as triggers, the system could deliver a brain-scrambling, heart-stopping jolt if the person attempted to walk through. This jolt would give the wearer an impression of walking into an invisible *Star Trek*-like force field.

Even today, researchers are experimenting with brain implants designed to alleviate a wide variety of conditions, including bouts of depression, Parkinson's disease, and autism, by delivering mild electrical shocks to specific areas of the brain.[9] The U.S. military's Defense Advanced Research Projects Agency (DARPA) already has demonstrated a noninvasive device for controlling airborne drones by using thought.[10] Combine these activities with the possible prisoner-control devices described above, give them all an additional 250-plus years of development, and there may be little to no need for a prison system. Whether such devices could be used to create human zombies under the control of the authorities remains to be seen, but it is a real possibility.

Considering the incredible destructive potential of modern weaponry, the Martian War will ironically be similar to the American Revolution in the selective way invaders used their power, as opposed to World War II, in which massive casualties were inflicted deliberately on civilians. In the American Revolution, the Howe brothers—one in charge of the British Army and the other in charge of the Royal Navy—felt that the war could be won with a display of overwhelming military might combined with a willingness to consider accommodations. They avoided inhumane or overly destructive actions that could enrage the colonists, in the belief that this approach would encourage the rebels to capitulate and amicably accept a postwar reestablishment of British rule. The Howe brothers viewed the rebels as fellow subjects, albeit misguided ones, rather than as foreign and evil or subhuman enemies.

Paradoxically, the Howes were not just tasked with leading Britain's army and navy, but also were officially appointed as the sole peace commissioners.[11]

On Mars, even with a desire on both sides to avoid massive destruction, the penalties for losing will be very high. For the colonists, the penalty for losing will be a loss of freedom in the conventional sense. However, given prisoner technology, it also may be a loss of free will.

The Nonlethal Alternatives

While the Martians definitely will prepare an array of lethal weaponry, they also will arm themselves with nonlethal choices. In today's world, these include low kinetic energy impact munitions (beanbags, rubber bullets, and sting grenades), chemical irritants (tear gas and pepper spray), tasers, various forms of blinding/disorienting lights, intense sound devices, electromagnetic devices (some of which can produce the intense perception of burning heat without actual physical harm), sticky foam that glues opponents to the floor, and so forth. Even in today's world with body-armor wearing, highly motivated soldiers, many of these nonlethal weapons are unlikely to be effective. On open ground, invaders would be wearing low-pressure resisting suits and using specialized breathing apparatus. They also might have this gear on them inside of tunnels. Wearing such equipment would eliminate pretty much any direct contact with chemical irritants.

This is not to say that nonlethal chemical weapons would have no value for Mars defenders. At some point, the invaders will need to rest and take off their gear. At that point knockout gases could be used to disable them. For example, in the 2002 Moscow Theater hostage crisis, Chechens held a theater full of innocent people hostage, vowing to blow them up if police attempted a rescue. Since the building was wired with explosives, the officials felt that commandos could not simply rush it. Instead, they flooded the theater with a chemical agent designed to knock out the terrorists so they would be unable to detonate their explosives. While all the terrorists were killed (mostly by bullets) when Russian special forces stormed the theater, dozens of innocent theatergoers also died from the effects of the chemical agent. Presumably, after many decades of further development, such agents could be refined to be less lethal.

Chemical agents or drugs could be delivered at night to unsuspecting invaders via insect-sized robots. These would be programmed to stealthily detect, latch onto, and inject their targets with the various agents, or they could inject agents into food or water. These agents could be deadly toxins and be used as a negotiating tool by the Martians if they were the only ones who had the antidote. Yes, insect-bots potentially could be defeated with mosquito nets, but then the bots also could be designed to shoot tiny needles short distances through the nets. Besides, even though nets treated with insecticides have reduced the incidence of malaria on Earth, they have not eliminated it.

Due to distance issues and the wide dispersion of energy beams, intense sound, light, and electromagnetic weapons would have limited use on open ground but could be overwhelming when used inside tunnels, where intense beams could be focused in a single direction and limited area. By the time of invasion, Mars would have thousands of miles of tunnels, and these could be equipped at key points with a combination of super-loud high-frequency and low-frequency sound devices along with high-intensity light sources, all designed to incapacitate human attackers. In this situation, ear protection, body armor, breathing apparatus, or pressure suits would provide little to no protection.

While sticky foam sounds ridiculous, trials with it have shown it can totally incapacitate individuals. In its initial form the foam was so sticky it would suffocate a person if it covered his face. The foam would also take hours to remove. Later generations of the product were reformulated to make it less lethal and easier to clean up. Used in an enclosed space, a sticky foam bomb would end a fight and make escape impossible.

After 250 further years of development, today's nonlethal weapons will seem medieval in retrospect. Given newer, more effective nonlethal weapons, there is a possibility that many of today's top lethal weapons might join poison gases, flamethrowers, and dumdum bullets on the list of abandoned materiel. On the other hand, given the consequences of failure, the eventual reality of Martian casualties (military and otherwise) and Martian military training using state-of-the-art virtual reality simulators, the Martian response to invasion could turn lethal very quickly.

NOTES

1. Kyle Mizokami, "Become a Super Sniper: DARPA Is Turning .50 Caliber Bullets into Guided Rounds," accessed June 6, 2019, https://nationalinterest.org/blog/buzz/become-super-sniper-darpa-turning-50-caliber-bullets-guided-rounds-27101.
2. Stephan Wilkinson, "The Goldilocks Fighter: The F6F Hellcat," accessed June 6, 2019, http://www.historynet.com/goldilocks-fighter-f6f-hellcat.htm.
3. For an excellent analysis of the interaction between technology and training with specific reference to the Battle of 73 Easting, see: Stephen Biddle, *Military Power: Explaining Victory and Defeat in Modern Battle* (Princeton, NJ: Princeton University Press, 2004), 132–49. See also Andrew Knighton, "The Battle of 73 Eastings—The Mother of All Battles?," accessed June 6, 2019, https://www.warhistoryonline.com/featured/battle-73-easting.html/2.

4. Mike Springer, "Learn How Richard Feynman Cracked the Safes with Atomic Secrets at Los Alamos," accessed June 6, 2019, http://www.openculture.com/2013/04/learn_how_richard_feynman_cracked_the_safes_with_atomic_secrets_at_los_alamos.html.

5. Atomic Heritage Foundation, "Espionage," accessed June 6, 2019, https://www.atomicheritage.org/history/espionage.

6. S. L. A. Marshall, *Men against Fire: The Problem of Battle Command* (Norman: University of Oklahoma Press, 2000).

7. Dave Grossman, *On Killing: The Psychological Cost of Learning to Kill in War and Society* (New York: Back Bay Books, 2009).

8. Anthony Robinson, Frank Hardisty, and George Chaplin, "Technology Trends—Smart Dust and Sensor Networks," accessed June 6, 2019, https://www.e-education.psu.edu/geog583/node/77.

9. Tim Sandle, "New Approach to Treating Depression with Brain Implants," accessed June 6, 2019, http://www.digitaljournal.com/tech-and-science/science/new-approach-to-treating-depression-with-brain-implants/article/538060.

10. Sputnik News, "DARPA's New Brain Chip Enables Telepathic Control of Drone Swarms," accessed June 6, 2019, https://sputniknews.com/military/201809071067847857-Brain-Chip-Fly-Drone-Swarm-Telepathically/.

11. Andrew Jackson O'Shaughnessy, *The Men Who Lost America: British Leadership, the American Revolution, and the Fate of the Empire* (New Haven, CT: Yale University Press, 2013).

Chapter Sixteen

Earth Invades Mars: A Strategic Overview

Assuming that the Earthlings could approach Mars without being detected and destroyed, they probably would focus their efforts on securing one or more of the space ports on the Martian surface. The space ports are a logical target for an initial landing, because they would have all that an invading force needed to establish a foothold on the planet, replenish supplies, and begin pacifying the local population. These space ports also will have direct access to surrounding power plants and sources of oxygen, two essentials for establishing a foothold on the planet.

Moreover, control of one or more space ports and the surrounding support infrastructure would provide a source of potential leverage over the hostile locals. In the most brutal form, the Earthlings could attempt to crush the resistance by cutting off rebels' ability to travel to other planets or survive on Mars without cooperating with the Earthlings. For this reason, control of these ports of entry and exit and their surrounding support facilities will be a critical objective for the initial landing force.

Once the Earthlings land, they must take full advantage of the initiative. In the first hours of the invasion, the Martian rebels would be shocked and might have difficulty communicating to their various rebel groups and coordinating an effective response to the invasion. These rebel groups eventually would be able to consolidate. But in the initial phases of the offensive, the Earthlings would have the advantage of more concentrated forces and a clear concept of the mission, and therefore would be able to dictate the scope and pace of military operations.

If all goes well during the first few hours of the assault landing, the Earthlings could quickly establish a security perimeter, secure all of the port facilities and the workers stationed there, cut communications out of the space port, and begin to covertly deploy a mix of robots and people to go out among the Martian population—all while minimizing casualties and collateral damage.

While the Earthlings might be tempted to consolidate their position or take a defensive posture, they will need to fully exploit their advantage of surprise during the opening phases of the invasion.[1] Failure to achieve any of these critical objectives risks the viability of the mission and may result in the Earthlings becoming trapped in the initial landing areas, ultimately locked in an unwinnable war.

Because of the crucial nature of these assignments, the Earthlings will be willing to use extreme violence from the moment they land. While a "shock and awe" strategy might appeal to some, the Earthlings would be wise to minimize their use of violence and instead try to limit their use of force and their impact on the Martian people and infrastructure. From a military standpoint, such an economy of force strategy would help maintain surprise, preserve the limited combat power of the Earthling forces, keep their tactics and capabilities a secret, and ensure that the critical port and power facilities would remain undamaged.

On the political side, such restraint would help establish the narrative that the Earthlings were not attempting to oppress or kill the Martian population and could be trusted to provide enlightened administration and authority over the rebellious planet. While the Earthlings will be seen by many Martians as unwanted and potentially dangerous aliens, the majority of the population will be unwilling to openly resist in the opening phases of the campaign. For Earth, it is vitally important to ensure that the majority of Martians do not fight back. The Earthlings do not need to truly win the hearts and minds of the Martians to achieve this goal, but they must not openly antagonize them and make armed resistance preferable to, or safer than, remaining neutral. While the Earthlings would have to find Martian rebels, fix their position long enough to plan a mission, and finish them with military means, each of these tasks would be easier in an environment that was less hostile to the Earthlings.

If history provides any insight into the realities of this contingency, this will be a very difficult task.

THE AMERICAN REVOLUTION REDUX?

Assuming that the Earthlings control one or more of the Martian space ports and have a secure base of operations, they then will focus their efforts in crushing the rebellion and reasserting their authority over Mars. While this campaign could take many different forms, the most likely and intriguing, again, is adopted from the American Revolution. The British had massive economic and military advantages over the American rebels, but were unable to ever force a decisive battle, win the support of the majority of the local population, or control enough of the countryside at any given time to actually assert their authority. While the British won battles, destroyed American cities, and pushed the Rebel army to the brink of collapse multiple times, ultimately they were unable to win.

Much like those Redcoats two and a half centuries before, the Earthlings may very well be able to control the ports and much of the urban infrastructure of their discontented colony. While necessary for resupply and transportation back to the home country/planet, this will prove insufficient for victory. In fact, the Earthlings easily could become bogged down in their own space port bases, just as the British did in the early years of the Revolution.

Even before the outbreak of open hostilities, the British realized that they were outnumbered, outgunned, and surrounded in the port of Boston. While not at war with the colonists, they feared that they could not win against a more numerous patriot populace that was well armed, if not formally organized and trained.

These tensions, combined with a mission to reassert the crown's authority, led British General Thomas Gage to order a raid on a rebel munitions storehouse at Concord, Massachusetts. Taking full advantage of their intelligence network, the colonists anticipated the British move and sent out riders, including Jimmy Dawes and Paul Revere, to warn their neighbors that "the British are coming."

The colonists responded quickly and were able to move their munitions and send a force of militiamen to meet the Redcoats at the Lexington village green on the morning of April 19, 1775. The British attempted to force the colonists to disperse peacefully, but a shot rang out, and the scene quickly devolved into chaos. While nobody knows who fired the resulting "shot heard 'round the world," the resulting firefight killed eight Americans and wounded one British soldier. While a militarily insignificant skirmish, the bloodshed at Lexington served as the spark that began the American War of Independence.

After quickly sweeping the colonial militia from Lexington, the British continued their march on Concord but were met by a larger force of patriots. While the British initially had the advantage in numbers, they were quickly overmatched by the American militiamen, who were swarming into the area. After a brief skirmish at North Bridge, the British realized that they would be unable to continue their search for weapons and began to retreat back to Boston. They were harassed by fire from the American militiamen for the remainder of their march and ultimately suffered 73 killed, 174 wounded, and 53 missing, casualties far greater than their American foe.[2]

In the days following the Battle of Lexington and Concord, some 15,000 patriots gathered around Boston. The British were surprised by their defeat as well as the ferocity of the American resistance. While they planned their next move, the British hunkered down in Boston, safe from an attack by the Americans, yet also unable to move out and take the war to their enemy. For almost the next year, the stalemate continued in what is today known as the Siege of Boston.

The most dramatic event during the siege was the British attack on Bunker and Breed's Hills on June 17, 1775. While the British were able to capture the rebels' positions, they did so at a terrible cost of over 1,000 killed, wounded, or missing.[3] Ultimately, the British were unable to break the siege and were forced to use their navy to evacuate their army to safer ports after the arrival of American heavy artillery made their positions untenable.

The British were able to escape potential entrapment and destruction in Boston but were unable to translate this additional tactical mobility into victory. Ultimately, they were able to push the Americans' Continental Army around in a series of campaigns but never able to defeat it decisively. Recognizing their inability to destroy the Continental Army in New England and the mid-Atlantic colonies, the British attempted to win political victories in the Carolinas, where support for the crown was generally greater.

Here, they met with mixed success. While they were able to win some of the hearts and minds of the locals, capture coastal cities such as Charleston, and inflict a major defeat on the Americans at Camden, they were never able to destroy the Continental Army. Much like they had experienced during their campaigns in the northern colonies, the British were able to win battles and capture key landmarks but were unable either to destroy the conventional army or win the support of the local populations.

Moreover, many of their pacification efforts proved heavy-handed and encouraged local irregular forces to rise and fight alongside the regular American troops. This inability to destroy the regular army combined with the increased fighting ability of the regulars resulted in the British defeats at Cowpens and King's Mountain. These twin losses led to the decision to retreat north to Virginia in hope of linking up with the British forces there and winning victories in that theater. Here again, the British were unsuccessful at achieving decisive military or political results. The final defeat came at Yorktown, Virginia, when a combination of American and French forces trapped General Charles Cornwallis's forces and forced them to surrender.

LESSONS LEARNED FROM THE BRITISH FAILURES IN THE COLONIES?

Failure to Rapidly Break out from the Landing Zones Cedes the Initiative to the Rebels

While the British would have numerous successes in subsequent Revolutionary War battles, their inability to break out from their base in Boston is very instructive. On the most basic level, it suggests that superior military technology and access to ports may not guarantee victory. Interestingly, the very ability to land in and secure ports may trap the Earthlings there and ultimately lead to their defeat. Much like the British in Boston, the Earthlings may be surprised when the Martians resist their efforts to move out from their bases and may quickly become trapped in the bases and faced with the prospect of an uneasy stalemate.

This uneasy stalemate will be a major benefit to the Martian rebels, as it will allow them to organize their resistance, prove their ability to resist an invasion, win support from the locals, and erode the combat power, logistics, and political support of the Earthlings. Time will work in favor of the rebels, and any delay in moving out of the initial landing zones is a victory for the resistance.

Breakout Battles May Be Unexpectedly Costly

The Earthling commanders, realizing that time is not on their side, will attempt a breakout from their bases once they have secured a perimeter, rested, and resupplied their forces. If met with a dedicated and organized resistance, they can expect to be faced with a series of difficult battles in and around the egress points from their bases.

Just as the geography of Boston prohibited easy egress from the port and into the surrounding countryside, the fact that space ports are likely to be connected to settlements by a few narrow passageways will make the paths out of the ports predictable, easy to defend, and easy to destroy. The result will be that much like the Battle of Bunker Hill, a ragtag force of rebels may inflict a surprising amount of damage to the Earthling forces by funneling their attack into a narrow and predictable avenue of approach. While the tunnels and transit points on Mars will be different from the geography of Boston and Charlestown, Massachusetts, the same military principle applies: When clustered into a killing zone, even the best military forces are vulnerable![4]

Winning a Series of Battles May Be Irrelevant

Even if the Earthlings could move out of their bases and engage in pitched battles with the Martian rebels, they may find it very difficult to achieve a lasting victory. Indeed, given a broad base of political support, the Martians should have little problem attracting recruits to their cause, even if they suffer significant losses in the subsequent battles.

Furthermore, the Martians always will have the option to retreat and cede ground to the Earthling invaders. Rather than being forced to defend any one location to the last man and risk losing their entire army, the Martians always can retreat and preserve their force. Much like the Continental Army allowed the British to temporarily capture New York, Philadelphia, Charleston, and other major cities, they could abandon these areas knowing that the British would be unwilling or unable to spread their forces out in order to occupy every major settlement or industrial center. Even if the Earthlings capture key geographic centers, they eventually will be faced with the choice of staying in these cities and giving up the initiative or abandoning them to pursue the rebel forces. The result will be that the Earthlings probably will never be able to completely destroy the Martian army, and winning battles and capturing points on a map will mean little in the long run.

Because of Distance and the Difficulty of Resupply, the Invading Armies Are Fragile

A related dynamic is that because of the extremely long supply lines, the invading force is perilously fragile. While the Earthlings will do their best to live off the Martian resources and recruit supporters from the local population, they cannot guarantee a reliable supply of basic goods like food, water, energy, and military equipment or replacements for the personnel and H-bots lost in battle. Because of this, they must be casualty averse. Because the Earthling commanders will do their best to minimize losses, they may be unable to act decisively and pay the costs needed to win for fear of losing too many troops in pursuit of victory. This factor, combined with the understandable political aversion back home to high losses, would make the job of any Earthling general extremely difficult. Although the invasion force may be impressive on paper, it will find it incredibly difficult to operate in a non-permissive environment millions of miles from home and may be surprising fragile despite its impressive combat power.

Stalemate Favors the Rebels

Given the immense political and logistical pressures, stalemate is the same as defeat for the Earthlings. This is an enormous advantage for the Martians as they do not have to win to achieve their political ends. Much like the American colonists realized, time is on the Martian side, and as long as they are not decisively defeated, time favors the defenders with the shorter supply lines and local support.

Given the Mix of Military, Geographic, and Political Factors, the Martians Almost Certainly Will Win

Given the mind-boggling combination of military, political, and geographical factors at play, it seems incredibly unlikely that the Earthlings could win. As discussed in a previous chapter, perhaps their best bet would be a population control strategy where they forced Martian civilians into camps, but this would carry its own set of political challenges and, if unsuccessful, would result in a genocide that would weaken political support on Earth while radicalizing the undecided Martians on the red planet.

While it is difficult to write this as a loyal Earthling, the most likely result of an invasion of Mars would be a Martian victory.

NOTES

1. For an instructive case study, consider the failure of the Allies to exploit their initial successes after the Anzio landings during World War II, see Carlo D'Este, *Fatal Decision: Anzio and the Battle for Rome* (New York: HarperCollins, 1991).
2. For a very readable account of the political climate surrounding the outbreak of war as well as the battles themselves, see generally George C. Daughan, *Lexington and Concord: The Battle Heard Round the World* (New York: W. W. Norton, 2018).
3. In fact, British General Clinton even referenced the ancient general, Pyrrhus of Epirus, when he stated that "few more such victories would have shortly put an end to British dominion in America." Henry Clinton, *The American Rebellion: Sir Henry Clinton's Narrative of His Campaigns, 1775–1782*, edited by William B. Willcox (New Haven, CT: Yale University Press, 1954), 19.
4. For an extreme example of troops being slaughtered in a killing zone, consider the ancient battle of Cannae fought in 216 BC. While ancient sources vary greatly as to the number of Roman dead, the vast majority of the Roman force of 86,000 men was killed, captured, or missing after being surrounded by Hannibal's much smaller force of Carthaginian and Gallic warriors. See: Mark Healy, *Cannae 216 BC: Hannibal Smashes Rome's Army* (Oxford: Osprey Publishing, 1994).

Chapter Seventeen

Physical and Cyber Counterattacks: Don't Mess with a Martian; He Might Throw Rocks at You

Given the difficulties of mounting a second invasion, repelling the first may be enough to win independence, but history shows how it takes more than winning a war of independence to fully establish a new republic. The eight-year American Revolution the colonists successfully won in 1783 was followed 19 years later by the War of 1812, fought once again with America's former colonial masters and sometimes regarded as a second war of independence. At the time, the British were at war with the French, and, as a result, interfered with American trade in order to prevent goods from reaching France. This interference included boarding American ships, kidnapping American sailors, and pressing them into service on British ships. The affronts did not stop at the water's edge. The British also were arming American Indian tribes in an attempt to limit America's westward expansion.

In terms of winning or losing territory, the War of 1812 accomplished nothing for either side, but it did elevate the former colonists to a more equal footing with the British.

Likewise, if Earth granted Mars its independence, there still would be issues, like who would control the exploitation of the mineral resources in outer space and whether Mars could still be used as a launch pad for Earthling efforts to do so. If Mars were viewed as subservient and unable to retaliate when bullied, Earthlings would remain as the overseers of outer space even though they may have relinquished direct control over Mars.

The threat of Earth's continued paternalistic attitude—or worse, another invasion attempt—would demand that the Martians be prepared to respond. While an invasion of Earth by Martians that ends in the Earth's total defeat might be a feel-good fantasy, its end point, if actually pursued, would almost certainly be grim for both sides.

A better strategy: Create and nurture double-mindedness in the Earthlings. This has a powerful influence on encouraging peacefulness. In this regard, generating cold-blooded fear is a useful tactic, such as the mutually assured destruction (MAD) policy pursued during the Cold War. Encouraging sympathy for one's cause and appealing to moral virtue also can help. In any event, some means would need to be found for establishing a balance of power.

There is generally some level of double-mindedness in any adversary. For example, at the start of the American Revolution, when members of Parliament could not restrain their enthusiasm for denigrating the character of the American rebels, a number of British army and navy officers refused to fight in America as a matter of conscience. On learning that his regiment was going to ship out to America, the Earl of Effingham resigned his commission as a matter of principle.

Even in the most patriotic of times, there is some double-mindedness present. Following the Japanese attack on Pearl Harbor, the United States reached a pinnacle of unification toward the goal of defeating Japan, and yet when put to a vote in the House of Representatives, a Republican from Montana, Jeannette Rankin, voted against declaring war. She previously had voted against America's entry into World War I and justified her vote against both wars on pacifist grounds. As a result, the votes have sullied her legacy to the present day.

For creating double-mindedness, weapons are useful mostly if not used. The Japanese attack on Pearl Harbor is a good example. With an economy about one-tenth the size of the United States', the Japanese knew they could not win an extended war. Admiral Isoroku Yamamoto, the architect of the Pearl Harbor attack, clearly stated to his superiors multiple times that once begun, he could not be victorious for more than the first 6 to 12 months of the war.[1] Hence, the Japanese strategy for winning depended on American double-mindedness causing them to capitulate after a few decisive Japanese victories. In reality, the bombs that fell on Pearl Harbor vaccinated the Americans against double-mindedness and led to the rallying cry, "Remember Pearl Harbor!"

THE CYBER-WARFARE CAPABILITY

Using information-based or cyber weapons is less risky than using physical ones. When used covertly, they can help for everything from gathering intelligence to blackmail, or even outright attacks resulting in serious monetary damage and loss of life (though loss of life may be counterproductive). Cyber weapons enjoy another significant advantage: Compared to physical weapons, they're inexpensive.

Prior to open conflict with Earth, Mars would lack a significant military but have a significant cyber-security force. In terms of digital capability and interconnectivity, Mars would be the most technologically sophisticated planet in the solar system, partly due to a necessity for survival, partly due to a local wealth of computer professionals, but also partly due to the fact that its computer systems would be built free of legacy issues. To support exploration and remote control of robots, the entire planet would be wirelessly connected via satellites. Mars also would be wirelessly connected with Earth, and while transmission delays would prohibit real-time transfers of information, it nonetheless would be possible to transmit digitized information such as internet content.

Wireless transmissions are more easily monitored than hardwired or fiber-optic systems, not to mention more susceptible to noise and interference. Given the amount of wireless data transfer on Mars, a significant design effort would be needed to prevent data loss from noise as well as unwanted, possibly nefarious hacking.

As an advantage over Earth, Mars would have almost the entire electromagnetic (EM) spectrum available for building its planetwide wireless systems. These would handle all forms of telecommunications, from entertainment to controlling billion-dollar mining operations on asteroids and possibly on Mars itself. There would be no need to dedicate valuable bandwidth in the EM spectrum for legacy applications like AM or FM radio stations.

On Mars, software-defined radio technology along with wide-bandwidth antennas would be commonly available for wirelessly connected digital devices. These technologies would enable communicating devices to change frequencies to less-used parts of the EM spectrum as needed to prevent data collisions. The technologies combined with a more openly available EM spectrum also would facilitate things like frequency hopping. Radios communicating with each other would "hop" from frequency to frequency in a seemingly random manner with the hopping determined by pseudorandom number generators (PRNG). Both the transmitter and receiver would use the same type PRNG set with the same seed value. The random number calculated by the PRNGs would determine the frequency setting of each radio. Each time a frequency hop was needed, the PRNGs of each unit would generate a new number to determine the new frequency, but since the PRNGs in each radio had the same initial value, the newly generated numbers would match, and the two radios would remain synchronized. Schemes of this type make it difficult to detect radio transmissions, let alone eavesdrop on their content. Of course, the same type system can also be used on Earth, but with fewer choices of frequencies available for hopping.

The Mars Cyber Corps (MCC) likely would be one of the solar system's most-talented cyber-warfare/security groups, given its unique experience with a high level of interconnectivity and widespread use of wireless communication/control systems used for critical functions. Hack into and shut down the U.S. electric grid for a few months, and its population would be in danger of starvation. Do the same thing to Mars's grid for a few hours, and its population would be in danger of asphyxiation. Let a hostile competitor hack into a Martian-controlled mining operation's robotics system, and it could shut down the operation after doing billions of dollars of damage. This could be used for blackmail or, if done surreptitiously, for putting competitors out of business.

Of course, the MCC would have limited resources compared to cyber-warfare/spying agencies such as the NSA, CIA, and various military cyber-warfare groups on Earth. However, like the conflict between David and Goliath, in cyber warfare, massive size and exhaustive resources are not required for doing devastating damage. Consider a cyber-warfare attack launched against a U.S. company by one of the smallest, poorest, and most isolated countries in the world, one that can barely feed its people: North Korea.

The 2014 Sony Pictures movie *The Interview*, depicting the leader of North Korea being assassinated by a bungling pair of journalists, was supposed to be hilarious but had a serious flaw: The leader of North Korea didn't laugh. Instead, after making numerous protests to no avail, North Korea launched a devastating cyber attack on Sony Pictures, pirating thousands of files and making public numerous embarrassing e-mails written by Sony Pictures employees. During the attack, a few computer-astute staffers hurriedly unplugged their computers while others watched in horror as their hard drives were wiped clean, subsequently bringing Sony Pictures to its knees. Compared to the North Koreans, the MCC would be far more sophisticated and capable.

Cyber-War Tactics against Physical Assets

In the event of hostilities with Earth, the MCC would need to hack or neutralize Earth-controlled observation satellites orbiting Mars. Simply destroying them could create thousands of pieces of space debris that could collide with and destroy future satellites. Disabling the satellites also might be undesirable if they were a useful part of Martian infrastructure—for example, as a GPS system. The best approach would be to take control of them surreptitiously, so that only Martian-approved information would be passed to Earth.

Taking control of the satellites orbiting Mars could be done from the surface if the MCC had access to the satellites' onboard software. If access were blocked, control or at least neutralization could be accomplished using miniature parasitic space probes (PSPs). These would be modeled after the activity of cordyceps fungi: The spores invade a host insect such as an ant, grow tiny tentacle-like fibers inside the ant's body, take over the ant's brain, cause the ant to climb to the highest position on a plant, and then clamp its jaws around a twig or leaf. The ant then dies, and a tiny mushroom sprouts from its body, spreading new spores into the wind with the goal of infecting more ants.

In 2013 the Chinese demonstrated that an Earth-orbiting satellite could "kidnap" a second satellite.[2] While not the same thing as taking over its "brain," certainly that would be a logical next step in development. To do so in a stealthy manner suggests that the aggressor-satellite would need to be miniaturized, and by the time Mars is well along in its colonization, miniature space probes used to economically explore asteroids for mining opportunities would be a commonplace reality.

PSPs would grapefruit-sized or smaller. Once launched from Mars's surface, they could maneuver and attach themselves to other satellites in orbit. This would allow them to bore into the interior of the host satellite, connect to its computer, and insert code in its software, or, if unable to do so, totally disable it without blowing it up. If the parasitic satellite succeeded in taking control of its host's brain, the parasite could use the host's radio transmitters to spread viruses into ground-based computer systems communicating with the satellite. Given Mars's low gravity well, PSPs could be launched easily and cheaply by the dozens.

Well before hostilities broke out, the MCC already would have infiltrated many of Earth's computer systems and created backdoors into them through which they could monitor information and/or disrupt Earth's systems. These could include military, manufacturing, financial transaction systems, telecommunications systems, transportation systems, and so forth. While this might seem unethical in peacetime, consider the fact that essentially every significantly industrialized country on Earth does this on a regular basis, both to enemies and allies alike. Not doing it, at least to some extent, would be naive.

For many decades, Martians would be highly dependent on Earth for critical supplies. These might include enriched uranium to fuel Martian reactors (at least until thorium-based LFTRs were available), spaceships, solar cells, machinery, medicines, and food. After the colony was established, it would take about 100 years, possibly more, for Martians to develop the agriculture, mining, and manufacturing capability to be self-sufficient, and even then it might be a better use of resources to import some of the desired products from Earth. During the truly vulnerable decades, it makes sense that Martians would be highly interested in knowing what Earthling leaders were thinking and in influencing Earth's policies toward Mars.

Due to the considerable time delays for two-way communication between Earth and Mars, normal forms of cyber attacks would be impractical. Nevertheless, hacking and cyber attacks could be done with the aid of a single computer system on Earth, communicating periodically with Mars. Such a system could use AI and only would need to be periodically supervised from Mars. However, the computer would not necessarily need to be a physical asset. It could exist as a type of super-virus distributed across numerous computer systems. The super-virus would wait patiently while opportunistically spreading itself into as many systems as possible. It likely would be incapable of entirely shutting down any single network, such as the power grid or financial system. Nevertheless, it could collect intelligence data or, if justified, create chaos by disrupting parts of numerous systems simultaneously.

Using an Earth-to-Mars wireless communication system would not be the only option available for hacking. Physical contact with a computer system would simplify hacking it. For example, what has been called "the most significant breach of U.S. military computers ever" occurred in 2008 when the U.S. Central Command computer network was breached.[3] A U.S. military member in the Middle East apparently found a thumb drive that had been dropped in a parking lot and inserted it into his laptop in hope of identifying its owner. Unfortunately, the drive had been planted deliberately by a foreign agent. As soon as the drive was inserted in the laptop, it uploaded a malware worm that proceeded to crawl around inside the Central Command Network undetected while searching for and transmitting valuable information to a foreign entity.

Using the SpaceX model, the numerous spacecraft landing on Mars will be serviced and fueled by Martians before returning to Earth. This would provide opportunities for Martians to hack the onboard computer systems and make surreptitious modifications. Eventually, the spacecraft's computer system would communicate with networks on Earth and, in the process, infect them. By altering the software on board a spacecraft, Martians conceivably could take control of it when it was making a return trip to Mars. If the spacecraft were used as a military transport for an invasion, Martians could misdirect it so that it would be hopelessly lost in outer space or simply could turn off all the life-support systems.

If the Earth-to-Mars transport were a large craft that stayed in space, shuttle spacecraft traveling back and forth between it and Mars could provide a physical link that could be used for altering the transport's computer software. Again, in the event of an invasion attempt, the transport's life support could be turned off or the shuttle carrying invaders be programmed to crash when radioed the command.

At the beginning of the invasion, uncertainty would prevail, causing Martians to hesitate before harming the invaders until after they had landed and made their intentions clear. Once those intentions were apparent, future would-be invaders onboard spacecraft with hacked computers would be in a precarious position—far more precarious than being fired on in outer space. If the first wave of an invasion stalled before reaching its objectives, a second wave could be wiped out before it landed. The Martians would need only to broadcast a triggering command to the spacecrafts' computers, sent at the speed of light in the general direction of the incoming spacecraft. This would be far easier than launching a drone spacecraft that would have to locate and attack the invading ship while avoiding its defensive capabilities.

Destroying a second wave of attackers with cyber attacks directed at invading spacecraft would be devastatingly effective for a while. But if the Earthlings were single-minded enough to continue after experiencing appalling losses, they could build new spacecraft with computers that had not been compromised and attempt yet another invasion. The delay in follow-up would be, at best, a few years and would allow the Martians time to prepare.

Deterring Future Attacks

Possible cyber attacks aimed at discouraging another invasion or in response to bullying incidents could be based on a strategy reminiscent of the 1951 movie *The Day the Earth Stood Still*, in which extraterrestrials, as a display of power, shut down Earth's electric systems for an hour. Such a shutdown would be possible for the Martians if they had backdoors to the right computer systems. Doing so for only an hour, however, would be just about impossible. Base-load plants, such as coal-fired or nuclear facilities, are not designed for rapid shutdowns and startups. It would take much more than an hour to restore their electrical power.

A better strategy might be to shut down Earth's GPS system for an hour. Given their PSP technology, the Martians could launch a swarm of them from the surface of Mars or from a spaceship on its return journey from Mars to Earth. These could attach themselves to Earth's various GPS satellites. In the best-case scenario, the PSPs would drill into and take control of the satellites, but they would not need to do so to disrupt them for an hour. The PSPs would need merely to jam the GPS signals.

Financial markets and banks use GPS to provide precise timing of transactions to prevent fraud (such as someone attempting to withdraw money simultaneously from two or more ATM machines). Precision timing via

GPS is not limited to the financial system. It also is used by the electric grid and, according to GPS.gov, "Some wireless services cannot operate without [GPS]."[4] So shutting down Earth's GPS likely would take down not only the financial markets and banks, but also the electric grid and cell phone systems.

But it would be even worse: By the time the Martian colony was well established, a majority, if not all, motor vehicles on Earth in industrial countries would be self-driven and using GPS for automated navigation. Shut down GPS, and one can only imagine the traffic jams. And the effect would not be limited to ground transportation. Airliners would not immediately fall from the sky, but keep in mind that a working GPS system allows more planes to fly in less space—an extremely dangerous condition without GPS. One way or another, air transportation also would be brought to a standstill. In addition, shutting down GPS could panic the U.S. military, which depends on it.

Although the disruption of Earth's GPS might last an hour, the effects could persist and involve loss of life and serious property damage, so attacking Earth's GPS might be like kicking a beehive. Still, the U.S. government typically has not responded to cyber attacks with equal or greater cyber attacks. In fact, it has barely responded. When the North Koreans attacked Sony Pictures, the U.S. response consisted of some mild sanctions and a lot of chest beating. Furthermore, the Obama administration made it clear at the time that America's nongovernmental entities were on their own for protecting their computer systems. The simple truth: The United States had far more to defend and lose in a cyber war than the Koreans. This resulted in considerable double-mindedness about where the red line is that would warrant a full-on cyber or physical response. In other words, Martians might be able to temporarily shut down the U.S. GPS system without a serious counterattack, but it would be risky. As for governments outside the United States, who knows what the response would be?

The Threat of Rock Throwing

While the Martians would have an initial advantage in a cyber war, Earthlings would have a clear advantage in nuclear weapons. So, at first glance, it appears that Earthlings potentially could fry the Martians without fearing a response. While nuclear power would be an important form of energy on Mars, if it were based on thorium, bomb making would be very unlikely. Even if uranium-based reactors were available along with a uranium supply, the magnitude of investment and infrastructure needed for nuclear bomb making would limit the size of the stockpile Martians could accumulate. Certainly, it would be small compared to the stockpile of nuclear weapons on Earth.

However, the Martians would not need nuclear bombs to make a serious physical attack against Earthlings. Instead, they could throw rocks—very, very large rocks. They could be made of durable materials like iron and nickel—in other words, metal asteroids that would impact the Earth with an explosion like a nuclear bomb, supposedly missing only the radioactive fallout—but maybe not. These do not need to be planet killers to do damage. A metal asteroid of a little under 20 meters in diameter would produce a Hiroshima-like blast when impacting Earth. Asteroids of this size would be difficult to detect as they approached.

As part of its terraforming efforts, Martians would know how to crash asteroids and comets onto the surface of Mars with enough accuracy to safely miss Martian colonies by a comfortable margin. Instead of creating miniaturized nuclear devices designed to release all their energy in milliseconds (a nuclear bomb), Martians instead would use a small-sized nuclear reactor—a technology certain to exist on Mars—coupled with an ion propulsion system that would release the reactor's energy—the equivalent of a nuclear bomb—slowly over the period of weeks or months to provide the thrust for moving an asteroid from its normal orbit to a collision path with a planet. At that point the target planet's gravity would take over and accelerate the asteroid downward until it either blew up in the atmosphere (typical of a stony meteor) or smashed into the surface (typical of a metal meteor). Either way, the damage could be considerable if it occurred in a populated area. Also, if the small-sized nuclear reactor rode the asteroid all the way down and blew up with the asteroid, there would be some radioactive fallout.

If installing small reactor-powered ion thrusters proved to be undesirable, the same effect could be accomplished with ordinary chemical reaction–based rockets. While not as elegant as moving asteroids with nuclear power, ordinary rocket fuel could do the job, especially when moving smaller asteroids.

When a mass is farther away from the Sun—in other words, in a higher orbit—it has more potential gravitational energy than the same mass in a lower orbit. If planets and asteroids were held magically in place by a giant tower extending from the Sun and an asteroid were released, it would fall toward the Sun with an ever-increasing velocity as the asteroid's potential gravitational energy was converted into kinetic energy. If a planet were attached to the tower far below the asteroid, the asteroid would smash into it at a very high speed. During the collision, the asteroid's substantial kinetic energy would be converted into heat in a matter of milliseconds, creating a massive explosion.

Since they are not attached to a tower, the celestial bodies in the solar system already are falling objects, except they are "falling" around the Sun. If a line is drawn from the center of the Sun to the center of a given celestial object, say an asteroid, it turns out that the asteroid has a very high velocity (called its tangential velocity) that acts perpendicular to the line (see figure 17.1), and this velocity keeps the asteroid from falling straight toward the Sun. Instead, the asteroid falls in a stable orbit around the Sun.

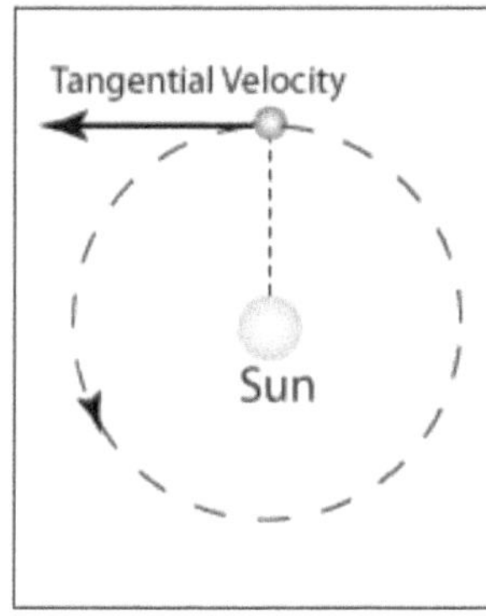

Figure 17.1. Tangential velocity of a celestial object in a circular orbit. *Source*: T. K. Rogers.

To make the asteroid fall toward a target planet in a lower orbit (lower meaning closer to the sun), we need to reduce the asteroid's tangential velocity. For the sake of a simplified explanation (it really is rocket science), let's assume the asteroid is in a circular orbit. If we reduce the asteroid's velocity just the right amount by using thrusters, the asteroid will shift into an elliptical orbit. Its furthest distance from the Sun will match its original distance, but its closest distance will match the target planet's distance from the Sun (see figure 17.2). If timed correctly, the asteroid will reach its lowest point at the same time the planet reaches the same location, and BOOM. Paradoxically, after slowing down the asteroid's original tangential velocity, the asteroid's overall speed actually will increase while moving toward the low point of its orbit due to the Sun's gravity. So the asteroid will slam into the planet under the influence of both the Sun's and the planet's gravities.

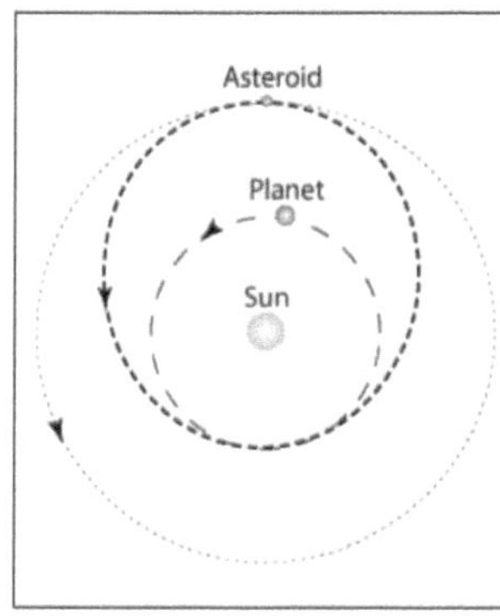

Figure 17.2. Changing an asteroid's orbit from circular to elliptical so that it can collide with a planet (elliptical orbit shown with heavy dashed line). *Source*: T. K. Rogers.

Precisely aiming the asteroid is problematic, but a miss of a few dozen miles from the center of a major metropolitan area is still going to be horrendous. In movies, given even a day's notice, an ICBM would be retargeted to blow up the asteroid in outer space, rendering it harmless. First, an ICBM would not be able to launch a nuclear warhead far enough into outer space, and second, a nuclear bomb doesn't have enough energy to render an incoming asteroid of significant size harmless, especially if it is a metal asteroid. Earth might have other means for protecting itself from asteroids, but detecting one in enough time would be difficult, especially for those still dangerous but significantly less than planet-killer sized ones.

Martians could combine "rock throwing" with an all-out cyber attack and a GPS shutdown. They also could overwhelm Earth's asteroid defense systems by launching numerous large balloons (inflated in outer space) and sending them traveling at high velocity toward Earth as decoys for the real rocks being thrown. If precision were needed for some targets, given a few years of preparation, Martians could build and launch numerous missiles with conventional warheads capable of hitting within a few feet of the target. A well-coordinated attack would knock out a significant amount of Earth's infrastructure, maybe enough to cause governments and economies to collapse, along with millions of deaths. Assuming Earthlings were smart enough to recognize the potential danger, they almost certainly would avoid serious future conflict with Mars so that the attacks would never happen.

After Mars had accomplished the first of the 4 C's of deterrence—capability for doing harm—making sure that Earthlings were smart enough to recognize it would depend on Martian success in the second C: communication. Martians would need to communicate their capability for attacking without giving away too many details or intimidating Earthlings so much that they would initiate a first strike. This could be done through diplomatic channels, press releases, and/or leaks. The communication would need to go both ways: Martians would need intelligence sources, such as an embassy, for collecting information on Earthlings.[5]

The capability for retaliation and degree of communication Martians need for deterrence would depend on their third C: calculation of risk versus cost. One of the key risks is a preemptive first strike from Earthlings that could wipe out the Martian ability to retaliate. For starters, if all ability to retaliate were located in one place, the risk would be high. The solution would be to disperse and hide both the retaliation capability and control similar to the way nuclear submarines were used during the Cold War. Considering the remote location of Mars from Earth and the Martians' space exploitation capability, dispersing retaliation capability and control throughout the vastness of the asteroid belt would be doable.

The fourth C—credibility—requires that capability be demonstrated when probed by real or threatened confrontation. In the wake of World War II, with the Russian army near the border of West Germany, the Americans positioned well-armed troops there, and while they may not have been able to repel an invasion, the loss of American lives would have triggered a serious response—a major deterrence greater than mere words could provide.

A new Cold War could develop between the Earth and Mars in which neither side has the ability to conquer the other at an acceptable cost. Thus, both sides would choose coexistence, as the Soviet Union and the United States—nations with very different languages and value systems—coexisted. Maybe, but it's more likely that the relationship would be similar to that of the newly formed United States and Great Britain, in which the former master's parental attitude toward its offspring evolved into the modern "special relationship," thanks to the mutual benefits of trade, commonality of language, and shared value systems. Let's face it: from Earth's perspective, the Mars colony would have been set up to facilitate Earth's defense against asteroid strikes and to provide access to valuable materials available on asteroids. Following the Martian rebellion, Earthlings and Martians alike would be motivated to patch things up.

NOTES

1. Hiroyuki Agawa, *The Reluctant Admiral: Yamamoto and the Imperial Navy* (New York: Kodansha America, 1979).
2. Jim Sciutto and Jennifer Rizzo, "War in Space: Kamikazes, Kidnapper Satellites and Lasers," accessed June 6, 2019, https://www.cnn.com/2016/11/29/politics/space-war-lasers-satellites-russia-china/index.html.
3. Jaikumar Vijayan, "Infected USB Drive Blamed for '08 Military Cyber Breach," accessed June 6, 2019, https://www.computerworld.com/article/2514879/security0/infected-usb-drive-blamed-for--08-military-cyber-breach.html.
4. NOAA, "Official U.S. Government Information about the Global Positioning System (GPS) and Related Topics," accessed June 6, 2019, www.GPS.gov.
5. For an overview on the basics of nuclear deterrence, see: Thomas C. Schelling, *Arms and Influence* (New Haven, CT: Yale University Press, 1966).

Chapter Eighteen

Negotiations and Aftermath: What to Do with the Tories?

If, as we predict, the Martians ultimately win their revolution against Earth, the strained relationship between the two planets' populations would be far from finished. A Martian victory would open up a new era of interplanetary diplomacy with a fascinating range of possibilities.

THE TEMPTATION TO DESTROY MARS

From the perspective of the Earthlings, continued conflict would be a bad idea. Once the Earthlings realized that they could not pacify the Martians, they may be tempted to completely annihilate Mars and end the problem once and for all. Despite the temptation for revenge or "ultimate victory," this would be an extremely high-risk strategy. If such a scorched-Mars strategy failed, the Martians could easily retaliate with their own scorched-Earth strategy.

This mutually assured destruction between the two planets should serve, at a minimum, to keep either side from attempting to eliminate the other and force each to explore other strategic choices than using WMD. Assuming that both sides were thinking rationally and were even slightly risk averse, they most likely would avoid this type of escalation, and Earth would be forced to accept the uncomfortable realities of a Martian victory.

While there may be some diehards on Earth arguing for a renewed conventional campaign against Mars, this also would be unlikely in the short term. Once Mars has demonstrated that it cannot be easily conquered and can exhaust the political will of the Earthlings, it would be very difficult to motivate the citizens of Earth for another campaign to win against a battle-hardened foe with significant physical advantages over Earth. If Earth accepts that it cannot win, anything other than an overt provocation by Mars probably will not cause Earth to restart the fighting. What is more likely is that Earth begrudgingly will choose to cuts its losses, accept Martian independence, and make peace.

Initially, the peacemaking process probably will begin with a cease-fire and exchange of prisoners. While such efforts would be laudable on humanitarian (or Martian) grounds, they will not result in a lasting peace. To move beyond a simple cease-fire, both sides would have to agree to a formal peace negotiation and hammer out a treaty acceptable to both sides.

This space diplomacy will be unprecedented and critically important for both sides. Despite the fact that the precise nature of the peace negotiations is unknowable at this point, diplomatic history and political science help make some predictions about the general course these discussions might take.

Formally Ending the War: An Interplanetary Negotiation

One of the most important issues facing the diplomats of Earth and Mars will be formally ending the fighting in a manner agreeable to both planets. While a cease-fire likely would be in place during the negotiations process, a significant potential for violence still would exist between the two planets. While the two sides were negotiating, both would want to maintain a relatively high degree of military readiness up to the point when the final treaty was signed. The reason for such tension is logically cold-blooded—the continued potential to inflict violence is a key part of any negotiation, the proverbial stick (threat of violence) to pair with the proverbial carrot (promise of future reward).

If at any time either side believes it is not getting a fair share of benefits at the peace table, it can always resume, or threaten to resume, the conflict. This potential to inflict future pain is a major source of strength, even for the defeated side, provided that it has at least some ability to carry out its threats.[1] Failure to maintain an adequate defensive capability could result in a draconian peace imposed on one side by the other. This was the situation which confronted imperial Germany at the end of World War I. The vanquished country faced starvation, revolution, and economic collapse, and was simply unable to credibly threaten organized resistance. Because of this, it was forced to accept the extremely punitive Versailles Peace Treaty, thus sowing the seeds of bitterness and ruin that would enable the rise of the Nazis in the coming decades.[2]

For the reasons already discussed, both sides should have the potential to use the threat of violence to promote an equitable peace. While crude, this capability may help both sides to extract what they believe is a fair deal and avoid aggressive bargaining tactics from the other side. Assuming that neither Earth nor Mars was completely ruined by the conflict, it would be unlikely that either would have such a military advantage as to extract a vastly disproportionate share of the spoils at the peace table.

The very fact that neither side could dominate the other using military means would make for interesting diplomacy; both sides would have to cooperate to find common ground for agreement while also giving up some concessions to the other. Despite the unprecedented nature of their work and the frustrations of delayed communications signals, diplomats on both sides would defer to an age-old system of bargaining, exchanging demands, and making counteroffers.

The negotiations would center on four major issues: 1) diplomatic recognition of Mars as an independent political entity, 2) the legal status of asteroids and other disputed territory, 3) armaments and arms control, and 4) people. In short: What is Mars's diplomatic status relative to Earth? Who controls the asteroids, other outposts, and the wealth and resources contained on them? How can Earth and Mars avoid a potentially disastrous arms race in the future? What happens to the soldiers, civilians, and prisoners of war who may be in a wide range of tenuous personal circumstances because of the recent conflict?

Diplomatic Recognition

Although it may seem quaint in the high-tech world of the future, formal diplomatic recognition will be a major point of negotiation. Since the Martians fought the war to achieve independence, this would be their first and most important point to settle. Earthlings may counter with a proposal for a system of Martian "home rule," in which the red planet had its own government structure but remained at least nominally loyal to Earth, not unlike the 1922 Irish Settlement or the British Commonwealth system. Such an arrangement should be a nonstarter for the Martians, since they had won enough military and political victories to bring Earth to the bargaining table.

For the rebels, anything less than full independence would seem like a betrayal of the principles of the revolution and would be politically unacceptable. Failure to win this basic concession from Earth would divide the Martians among themselves and thus make it incredibly difficult for any Martian diplomat to sign a peace treaty. If they did accept a treaty that only provided for limited sovereignty, it is entirely possible that Mars would face an internal struggle between pro-treaty and anti-treaty forces, similar to the 1922–23 Irish Civil War. While the Irish pro-treaty faction was happy to achieve limited independence and was fearful of Great Britain resuming the conflict, the resulting internecine conflict sowed bitter division within Ireland and led to tragic fratricide among the Irish rebel leaders.[3] The desire to avoid such divisions, along with a relatively strong bargaining position with Earth, should motivate the Martian negotiators to act in concert and insist on nothing short of full independence.

Furthermore, full independence and recognition would be the only way to actually solve the underlying issues that provoked the war in the first place. Since it began in large part because of the great difficulties of governing a planet over the vast distances of space, the dynamics of the one-way mission, and the conflicts between preserving versus terraforming Mars, it would solve little to go back to a system that simply provided a greater degree of autonomy to Mars. While Earth might attempt to dissuade the fears and claim that it would act in good faith if allowed to continue to exert control, it seems unlikely the Martians would take these promises seriously after their victorious struggle.

The result is that Mars, like the United States after the American Revolution, likely would win full independence from Earth. Earth might not like this outcome, but ultimately it would have very little bargaining leverage once the peace talks had begun. The Martian rebels thus would be granted their independence and sovereignty.

Territory

While Mars would insist on (and almost certainly win) full independence, significant territorial issues will remain. It would be relatively simple to declare that the Martians get control of the surface of their planet; the key sticking point will be the bases and infrastructure that Mars has established for asteroid mining, early warning, and deep-space exploration. Assuming that these operations are profitable and have been established with funding and logistical support from Earth, it will bring up a very real debate regarding who owns them and who gets the profits from these outposts.

This is a difficult issue, and both sides will make legitimate claims to ownership of the outposts. After giving up on the issue of Martian sovereignty and diplomatic recognition, the Earthling diplomats probably will choose to take a much firmer stance on asteroid holdings. Earthlings could claim justifiably that since their research, funding, and support went into establishing the outposts, they are the rightful owners. Moreover, the Earthlings could assert that since they are the biggest consumer of the goods from these outer settlements, they have a profound economic and political motivation to recoup their investments in these operations.

Mars might argue to the contrary that the success of these outposts was directly attributable to the hard work and sacrifices of the Martian people and that they should be allowed to continue to operate these facilities to ensure their continued success. The Martians' technical expertise at running the operations, combined with their advantageous physical location, would provide them with significant bargaining leverage, because cutting them out of the process would interrupt the flow of goods to Earth and incur significant economic costs in the short run.

Since there will be no established interspace court system for settling this dispute, diplomats from both sides will need to engage in multiple rounds of give and take to ensure that the issue does not derail the peace process. What most likely would happen is that the two sides agree to divide the outlying settlements or share the resulting profits in a mutually agreeable manner.

Perhaps one pathway to pleasing both sides would be that Earth would be given exclusive control of these mining and early-warning facilities for a set period of time before ultimately returning them to Mars. Such an arrangement would be akin to the British control of Hong Kong after the Opium Wars or the American control of the Panama Canal Zone. Such an arrangement would allow both sides to claim a diplomatic and economic victory. Earthlings could declare that for the foreseeable future the rights of their investors were protected, while Martians could take comfort in knowing that time was on their side, and they would eventually be able to claim what was rightfully theirs. Alternatively, Earth might accept a payoff from Mars to help satisfy its investors but leave the actual ownership of these settlements to Mars. Given a relatively equitable division, both sides could state that they secured a share of the profitable outposts and kept this issue from disrupting the larger peace negotiations.

Whatever the final outcome, it would seem likely that the asteroids and their products would continue to be a contentious issue even after the peace treaty was signed. The issue could provoke another conflict later, much like the lingering matters that led to the War of 1812 after the American Revolution. In the short term, though, both sides probably would focus their efforts on establishing a trading partnership to maximize profits from the asteroids and provide for the continued supply of resources; both begrudgingly would accept the fact that neither side has a monopoly on these settlements or exclusive ownership of outer space.

On the issue of early warning outposts to detect meteorite strikes, Earth will be largely at the mercy of Mars. Mars could choose to deny this critical information to Earth but at the risk of derailing the peace process. Such an aggressive tactic probably would backfire because it would seem unduly harsh, but it also would forgo the potential for profit by selling the information to Earth—or the potential to build goodwill by giving it free of charge. The result is that any peace deal probably would result in Earth retaining at least some access to the early-warning system that had been a key impetus for accelerating Martian settlement in the first place.

Armaments

Another issue likely to be discussed is disarmament. While both sides would have powerful incentives to limit the potential for future violence, there would be no good way to ensure effective arms-control measures. The same technology used for peaceful purposes such as terraforming and asteroid mining also could be employed to crash a meteor into the other planet. This potentially dual-use technology would be so essential for the continued development of Mars and so difficult to ban or monitor that it would be virtually impossible to control. Unlike the distinction between nuclear energy and nuclear weapons, there would be no means for delineating between peaceful and military purposes.[4] If you can redirect an asteroid for peaceful purposes, you can redirect the same asteroid for nefarious purposes.

The result will be that the best the negotiators would be able to do on the issue of WMD is to make relatively meaningless pledges on no first use, or to allow inspectors to monitor equipment used to mine asteroids or for transporting objects from one planet to another.

In such a system, cheating is always possible, and the best either side could hope for is that the threat of mutually assured destruction keeps the two planets from destroying each other long enough for feelings of trust and rapprochement to make the prospect of planetary annihilation an unthinkable taboo. Although the potential always would remain, it would be at least conceivable that over time, leaders on both sides eventually would stop thinking in terms of using these devices destructively, and this would further strengthen the security of both planets and their people.[5]

On smaller weapon systems such as H-bots and warbots, there also is little cause for optimism. The ability to program, 3-D print, and deploy civilian robots makes it possible to switch this technology to the production of weapons. Here again, pledges of goodwill and an inspection regime might help calm fears, but the potential for cheating and being taken advantage of by the other side will make it very difficult to achieve lasting arms control.

While arms control likely would fail, it does not mean that future conflict is preordained. The costly experience of the recent war, the physical separation between the two planets, and the potential for planetary destruction may lead politicians to conclude that there are few incentives for fighting in the future. If both sides understand that conflict would be costly, protracted, and possibly mutually destructive, then it is possible that the solar system might have little reason to go to war in the future, even if arms-control measures fail.[6]

People

Perhaps the most important issue for the diplomats to settle will be the fate of people. The war inevitably will leave many from both Earth and Mars in ambiguous political and legal states: political prisoners, prisoners of war, spies, stateless (or planetless) persons, and loyalists to Earth living on a newly independent Mars. It will be the job of the diplomats to attempt to untangle this complex human web, and failure to properly protect, punish, and provide for the needs of these people may very well be the enduring legacy of the conflict.

Political Prisoners

Because of the intensely political nature of the war, there potentially will be thousands of people arrested and detained for their affiliations. These prisoners can run the gamut from rebel leaders to propagandists to political dissidents. While the conflict on Mars would provide ample cause to hold them, once the war is concluded there will be a push for their release.

Although both sides may be reticent to set free these troublemakers, once the fighting is over they quickly will become a political liability. Holding them will be counterproductive, as they quickly could become symbols of oppression. If these political prisoners achieve widespread fame and support, they may act as an impediment to the peace process. Both sides probably will agree to exchange their captives and turn them over to the lawful authorities on the other side once a peace treaty has been signed.

Once the government flips, these newly freed political prisoners could gain the status of heroes among their respective sides, as they suffered for their cause. Some undoubtedly will use their martyr status to enter the realm of politics, as Nelson Mandela did after the collapse of the apartheid government in South Africa.

The result is that Earth might be forced to conduct diplomacy with its former prisoners, an unpleasant fact indeed. Despite the awkward nature of this outcome, the alternative of holding them and furthering their status as victims would be worse than simply setting them free and allowing them to run for office in their new government.

Prisoners of War

While the fighting on Mars will be fought with a large number of robots, there will be prisoners of war to deal with as well. Under current international law, combatants captured without proper identification or a uniform could be treated as spies and executed. Many of the Martian rebels might fit into this category due to the nature of the guerrilla wafare and the lack of recognition and legitimacy to the revolution. Because of the distinction, the Earthlings might be tempted to kill them in mass. Such a barbaric policy would undermine peace efforts if they refused to grant prisoner of war status to Martian fighters based on this arcane technicality.

Moreover, such a draconian punishment might lead to the unintended consequence of Martian reprisals on captured Earthlings and might cause the Martian resistance to fight harder for fear of death if caught. Failure to properly treat prisoners of war has led to countless incidents of reprisal killings and brutal combat, none more deadly that that experienced by the Nazis and Soviets during World War II. Because the Soviet Union was not a party to any of the conventions outlining the rules of war and the proper treatment of prisoners, the Nazis began a brutal series of murders of Soviet prisoners. The Soviets responded in kind, and ultimately millions of POWs were killed in this brutal tit-for-tat race to the bottom. Ultimately, only the unconditional surrender of Germany ended this bloodletting, and the memories drove suspicion and hatred for decades.

This brutality could bleed over into the peace negotiations if either side believed its prisoners were being treated unfairly. The offended party could threaten reprisals or simply refuse to negotiate. It would benefit each side to have a fair number of the other's soldiers in its possession, much like hostages during ancient and medieval times. Both sides may have reservations about releasing the captives, but negotiators probably will agree to a prisoner exchange as a means of furthering the peace process and avoiding a vicious cycle of reprisals.

Spies and Saboteurs

As discussed in previous chapters, spies and saboteurs probably will have a disproportionate influence on the Martian conflict–gathering intelligence, running false-flag operations, destroying critical infrastructure, and the like. This could lead both sides to severely punish suspected spies to make an example out of them and, interestingly, the law probably will be on their side. Under current international law, spies and saboteurs do not enjoy the same legal protections as members of the uniformed military, and they can be charged with espionage and other crimes against the state and executed. While both Mars and Earth may find rapid execution as the most expedient way of dealing with captured spies, any intelligence assets still living at the end of the conflict may become a key part of the peace process. Neither side will want to admit to the nature and scope of their covert activities, but they probably will spend significant political capital behind the scenes to ensure the safe return of their intelligence assets. To this end, discussions regarding the release of this special class of prisoners likely will be conducted in secret, far away from reporters and diplomatic observers.

Stateless Persons

An altogether separate issue will involve people trapped in limbo without a clear planetary identity. These could include citizens from either Earth or Mars who were visiting the other planet and were trapped there during the fighting, or those in transit between one planet and another. This is a common problem after major wars on Earth, where people lose their statehood or are not allowed to return to their country of origin for geographical or political reasons. While refugees and stateless people are a major humanitarian concern on Earth, they are an even greater potential issue in outer space because of the great distances between the two planets, the costs of transport, and the risks to their physical well-being from being forced to quickly leave one planet (or turn their space ship around) and travel to the other.

Although diplomats may use these people as political bargaining chips, the probable result is an uncomfortable and uncertain future for them. Much like the boat people after the Vietnam War or European Jews after the Holocaust, these vulnerable populations may find little help or support on either planet.

The very fact that the people are in political limbo will make it hard to negotiate on their behalf as they are not part of either side's political constituency. Without representation at the peace table, these people will have no bargaining leverage, and their fate will rest on the benevolence of others. While enlightened goodwill might protect them, these stateless people will have no guarantee of a satisfactory resolution to their plight. Representatives from Earth and Mars possibly could make some token resettlement efforts but will focus the bulk of their time and political capital in solving other issues of higher priority to their own political bases.

If the people cannot be integrated back into their respective home planets, they could wander around outer space in their own self-contained spaceship, much like the boat people of Vietnam. Similarly, they could attempt to settle on another planet or moon in the solar system, much like the Jews after the Holocaust. Both of these possibilities are intriguing for interplanetary relations and would impose significant personal and societal costs. Unfortunately, due to the extreme difficulty and expense involved in space travel and colonization, neither of these options is likely.

Loyalists

While the lonely limbo of the stateless persons would be a serious issue for the two planets, it would pale in comparison to the much bigger issue of dealing with the loyalist population on Mars. As previously discussed, not all of the citizens of Mars will support the revolt, and others will slowly lose their faith in the Martian cause. For the Martian rebels to win, they must maintain the support of the population, but they cannot win it over in its entirety. A significant percentage of the Martian population will remain loyal to Earth.

These loyalists will come in several forms, ranging from vocal critics of the Martian revolt who risked everything to demonstrate their allegiance to the mother planet to the silent supporters of Earth who remained loyal but kept their political feelings a closely guarded secret. No matter how vocal they were during the conflict, both of these groups pose a problem for the new government of Mars.

For obvious reasons, Mars will want to remove these elements from their planet to ensure that those who stay are willing participants in the task of building a new society with its own independent identity. Yet despite this clear goal, the process of purging the planet of loyalists almost certainly will be messy and could derail the peace process.

Many of the die-hard loyalists probably would be content to be released from Martian control and to leave the planet, but this would not settle the issue. They would want physical protection to ensure that they are not exposed and attacked prior to their departure, and they would make legal claims of compensation for property left behind or confiscated by the Martian authorities. These demands will present the Martian negotiators with a difficult choice: Do you accommodate the demands of the loyalists as a matter of political expediency, or is it worth risking the broader peace process to punish them?

Both options would have their appeal; both would carry risks. Providing safe passage off the planet and repaying the erstwhile citizens for their lost property would send the message to Earth that Mars was confident in its own position and was trying to act as an honest broker respecting the rights of all people. This choice would not necessarily play well on Mars, as it could lead Martian patriots to believe that the interests of the loyalists were being prioritized over those who had risked everything for the revolution. To appease the Martian political base, it would be tempting to strip the loyalists of their rights and publicly exile them from the planet.

To avoid a massive humanitarian crisis that could poison the relationship between the two planets, the Martians would be wise to adopt a middle course: provide physical protection for the outcast loyalists but only make a token effort to restore their property and compensate them for damages.

This would avoid the potential for reprisal killings against the loyalist exiles, which could lead to further violence, but would appease most of the Martian patriots who wanted to rid themselves of the traitors within their midst. Failure to do so could lead to a grassroots genocide, not unlike that witnessed during the partition of Greece and the Ottoman Empire or the partition between India and Pakistan, two incidents that resulted in the deaths of hundreds of thousands of people and has poisoned the relations between these countries to the present day.

The less vocal loyalists also would create a problem for the Martians. Because these people were more covert in their loyalties and probably more numerous, they would be harder to identify and separate from the rest of Martian society. Efforts to find and punish them would have the potential to get nasty, especially if the Martian authorities relied on informers to identify silent loyalists. This easily could lead to a nationwide crisis, with neighbors informing on each other to settle old scores as well as actually rooting out traitors. In such a paranoid domestic climate, it would become incredibly difficult to determine truth from lies, and Martian society could devolve into chaos.

Even if it were possible to untangle this complex web of identities, Martian politicians would be wise to forgo a systematic attempt to bring all of the loyalists to justice. It would be preferable simply to make it a government policy that all but the most public loyalists were free to remain as long as they were law-abiding citizens of Mars and recognized the authority of the new government. This process would allow the Martians to heal the social wounds of the recent war but would motivate many of the disloyal members to silently slip away. This was the tactic adopted by the Americans after the Revolution; and, as a result of this benevolent policy, many Tories fled to Canada or England, many chose to stay and contribute to the new nation, and the country was spared a protracted and bitter witch hunt to root out traitors from within.

Unfortunately for the Martian Tories, they could not simply flee to a "Canada" on Mars itself. Instead, when faced with the expense, time, and biological barriers inherent in returning to Earth, many would choose to remain on Mars, accept the new political arrangement, and integrate back into society.

This decision will be made even easier for all but the most recent immigrants to Mars, as they will have nowhere to go. They will not have a social support network back on Earth. They will not have a job waiting for them. They will not be welcomed, and they might even be viewed as outsiders if they returned to Earth. This harsh reality will make staying on Mars and putting aside politics the easy choice, not the difficult one.

While some feelings of mistrust may linger, the Martians would be wise to accept these people back into society. Such an enlightened policy would be a benefit to the newly independent Mars as it would prevent a "brain drain" of talented people from leaving and would allow the Martian economy to rebuild more quickly thanks to the efforts of these silent dissenters. Such a policy also would have the added benefit of signaling peaceful intent to the Earthlings; it would not be an impediment to peace in the short term or a source of friction in the future.

The Importance of Diplomacy and Perception(s)

This desire for peace does not end the troubled relationship between the two planets. Even if both sides are genuinely interested in making a lasting peace, serious new questions will arise about the new relationship that will take years to understand fully. Will an uneasy stalemate exist? Will the two planets engage in arms races and security competition, each biding its time for an opportunity to resume the war? Will they attempt to repair their strained relations and seek new opportunities for trade, cooperation, and peaceful coexistence?

All these questions are difficult to answer and would depend on the fairness of the peace treaty and how the leaders and populations on both planets view each other. If the peace treaty ends up as the interplanetary version of the Treaty of Versailles, it will lose legitimacy and sow the seeds of future conflict. But a just peace—with both sides making some concessions, balancing economic and military power, and protecting their core interests—could guarantee peace and cooperation for generations.

The mutually assured destruction dynamic should help ensure that the peace is acceptable to both sides. Because politicians on Earth and Mars will know that the other side can always walk away from the negotiations, they will be incentivized to never ask for more than what the other side could be reasonably expected to give up. If harsh reparations are sought, the talks will break down and the possibility for mass destruction again will be in play. This balance of terror should make both sides cautious and force them to forgo any attempt to win a draconian peace. The most likely result is that neither side can "win" at the peace table. An interplanetary Treaty of Versailles is not probable because of this latent potential for destruction, and both sides will have to compromise to produce a settlement that is both mutually agreeable and durable.

Much also will depend on whether the Earthlings see Martians as their departed brothers and sisters or a hostile alien race. If, as the British did, they assume that, despite the war, they have more in common with their departed colonies than they have differences, then the prospects for rapprochement should be significantly greater. If they feel embittered and resent their departed colonists, then they are more likely to harbor ill will and a desire for revenge for generations to come.

Unfortunately, these perceptions will be made even more complex by the inevitable use of propaganda and racism on both sides. As previously stated, the physical differences between Earthlings and Martians could engender feelings of xenophobia and racism, two emotions that would be exploited during the war to motivate each side to fight and kill the "other" group. Even in the future, humans are unlikely to change their minds once they have been made up, and old prejudices will die hard.

One possible pathway to changing the perceptions of those on Earth would be fear of an outside threat. In the early days of the Cold War, the Western allies recognized the need to rehabilitate their relationships with West Germany and Japan and quickly made efforts to become friendly with their recent enemies. Despite the fact that they had just ended the most destructive war in modern history, the fear of Soviet expansion compelled the Americans and British to rapidly change their perceptions of their former foes and find common cause against the outside threat.[7]

And there *is* an outside threat that may provide common cause for Earthlings and Martians—flying rocks! The same fear of asteroids that inspired Earthlings to accelerate their efforts to colonize Mars in the first place will still exist. Earth and Mars will remain vulnerable to these events, and this shared sense of threat might help heal the wounds of war. Because Mars still would be in an advantageous position to detect and respond to meteors, Earth probably would make the first move to thaw tensions. If Mars were receptive to these overtures, even out of profit-driven motives, it would be possible to quickly reestablish a cooperative relationship based on the exchange of information. Over time, cooperation on other, more contentious issues would become easier as both sides demonstrated their reliability and trustworthiness. This mutually beneficial relationship might not be the same as friendship, but it eventually should lessen fear, racism, and rivalry between the two planets' populations.

Hopefully, Earth and Mars would be able to peacefully coexist, or the next time Earth might end up with something far worse than a Hollywood-style Martian invasion.

NOTES

1. Fred Charles Ikle, *Every War Must End: Second Revised Edition* (New York: Columbia University Press, 2005).
2. Margaret MacMillan, *Paris 1919: Six Months That Changed the World* (New York: Random House, 2002).

3. Tim Pat Coogan and George Morrison, *The Irish Civil War* (Boulder, CO: Roberts Rinehart Publishers, 1998).

4. Seumas Miller, *Dual Use Science and Technology, Ethics and Weapons of Mass Destruction* (Cham, Switzerland: Springer International Publishers, 2018).

5. Nina Tannenwald, *The Nuclear Taboo: The United States and the Non-Use of Nuclear Weapons since 1945* (Cambridge: Cambridge University Press, 2007).

6. Robert Jervis, "Cooperation under the Security Dilemma," *World Politics* 30, no. 2 (January 1978): pp. 167–74.

7. Patrick Thaddeus Jackson, *Civilizing the Enemy: German Reconstruction and the Invention of the West*. (Ann Arbor: University of Michigan Press, 2006).

Conclusions

Our previous chapters beg the question, "Are we proposing that history repeats itself?" And, the answer is no . . . but also yes. Just considering differences in technology alone, a new event cannot be identical to a previous one. In addition, it's impossible to duplicate the countless number of details from the first event. Nonetheless, there are patterns. History is about people, and while things like technology and the information it's based on evolve at a dizzying pace, people evolve very slowly. Presented with various types of situations, people follow patterns. Often these patterns are based on emotion and may be irrational. But even irrational responses can be predictable.

Irrational behavior, it should be noted, is not necessarily bad. If people always followed their heads instead of their hearts, a lot of beneficial things would never get done. The facts required for a fully rational decision often are not available until after the decision is made. Certainly, irrational behavior (sometimes called risk taking) will be involved for people to abandon a lush planet with abundant supplies of water, to possibly lose not only their life savings but their lives, to relocate on an otherwise lifeless desert where one cannot even breathe normally.

The narrative and the many predictions contained in this book rest on the robustness of our assumption that two major historical patterns will occur with respect to Mars: First, if a vast new territory becomes economically reachable, investors and speculators will seek to profit from it; second, if the technology exists to make the territory habitable and expandable, settlers will flock to it in spite of the personal risks.

With regard to the first critical pattern, we have needed to establish that reaching Mars in an economical enough manner for colonization is within the reach of humanity's engineering capabilities. Zubrin's Mars direct plan, in which fuel for a return trip is manufactured on Mars, combined with the

SpaceX plan for building and reusing large-scale spacecraft, indicates that Mars will become reachable on a large scale. Yes, it is rocket science and expensive, but humanity already has spent the money to build rockets that have sent manned spacecraft to the Moon and various unmanned spacecraft to the far reaches of the solar system.

We also have needed to establish that investors and speculators will have a way to profit from Mars. Here, the huge possible payoffs from asteroid mining (supported from Mars) or possibly mining on Mars itself, along with revenue collected from transportation fees and contracts to support scientific exploration, appear to be sufficient to attract the needed investors and speculators. Again, there is an element of irrationality here, but being the first nation, corporation, or billionaire to legitimately enable the exploration of Mars or establish a colony on it is an achievement of historic proportions. It goes way beyond wealth, power, and ordinary prestige, which by comparison are commonplace. The possibility of realizing such an achievement is worth taking considerable risks. Once the thought of such an accomplishment has been planted, the risks and dangers involved will take on a romantic glow.

The second critical pattern depends somewhat on the extent to which Mars can be terraformed into an Earth-like environment. Based on a significant amount of information, we have assumed it can. Indeed, we think that issues regarding terraforming could be the final ones that trigger the Martian revolt. Honestly, there are only two logical reasons to go to the trouble of settling Mars: 1) to study it scientifically or 2) to exploit it. And these reasons can conflict. Scientific study doesn't require many people, and they are likely the type who would be content living in relatively confined quarters on another planet, considering the high possibility of making a significant scientific discovery. They will want to keep the planet as uncontaminated and unchanged as possible to preserve scientific evidence. Exploitation, on the other hand, requires change. It attracts a crowd, and while they might be willing to endure hardship at the beginning, their longer-term goal, if not for themselves then certainly for their progeny, would be greater wealth and comfort—the type that terraforming could help produce. After all, it is nice to be able to walk around one's planet and have enough surface water to settle some of the dust, not to mention have enough oxygen and atmospheric pressure to breathe. Unfortunately, the changes required to produce such niceties can mess up the collection of pristine scientific data.

A recent study casts doubt on the possibility of terraforming Mars, suggesting that such an undertaking is beyond humanity's capability. According to Bruce Jakosky of the University of Colorado, Boulder, and Christopher Edwards of Northern Arizona University, there is not enough CO_2 in the polar caps, regolith, and rocks to thicken Mars's atmosphere enough to have an atmospheric pressure anywhere near the desired level.[1] This would dash all hope of having a breathable atmosphere (in which most of the CO_2 was converted to oxygen by plants) or significant amounts of liquid surface water. Hence, our proposal that the atmospheric pressure will be raised to around five psi and the process of making it breathable will have begun in about 250 years may be completely wrong. But then, scientific knowledge about complex issues such as terraforming tends to be a moving target that can change radically over time.

Those of us who grew up in the 1950s knew that dinosaurs were giant cold-blooded reptiles. Today we think at least some of them may have been warm-blooded birdlike creatures covered with a type of feathers.[2] Even though dinosaur fossils have been collected and analyzed around the world by thousands of researchers for around 200 years, a few extra data points have radically changed our point of view. Considering that the surface of Mars has barely been explored, there are a lot of possibilities for finding new data that could favor terraforming.

In addition, dire scientific predictions of future conditions based on partial data and extrapolation often have been proven wrong when history finally catches up with the predictions. In his famous 1968 book, biologist Paul Ehrlich predicted humanity would experience widespread starvation in the 1970s, including millions of starvation deaths in the United States; that India was doomed; and even England might not exist by the year 2000.[3] Now, years after the predictions were supposed to have happened, none of them has and none seems eminent. History shines a bright light when it eventually illuminates the reality (or unreality) of predictions.

Finally, there are the unforeseen effects of new technology on supposedly insurmountable problems such as terraforming. As far back as 1919 and every few years since, experts have predicted that U.S. oil production was about to peak and then irreversibly decline. Generally, the time span for running out has been on the order of 25 years, sometimes less. Yet today, thanks to a significant development in technology, fracking, the United States is experiencing a major boom in oil and especially in natural gas production. Certainly, the U.S. oil supply eventually will run out, but it looks like it will not do so any time soon. Given all the above arguments, we are inclined to side with the many visionaries who believe that Mars can be terraformed, even though there are some well-informed scientists who are doubters.

While helpful, terraforming is not a necessary condition for creating and maintaining a significant human colony on Mars—a colony with the rocket fuel-making and manufacturing capability to become independent of Earth and evolve into the solar system's primary spacefaring planet. The critical element is the creation of an abundant energy supply that has a high return on equity. There are many reasons for this requirement, but creating the pressure and oxygen content required for breathing inside enclosed human habitats tops the list. Unlike the vast areas of forest on Earth, plants growing in greenhouses on Mars will have a highly limited amount of area, likely not enough to make sufficient oxygen for humans to breathe, certainly not in the early days of colonization and maybe never. In addition, plants cannot raise the atmospheric pressure inside human habitats. Much of the oxygen and all of the atmospheric pressure required for breathing will need to be produced by energy-intensive processes. From an energy consumption standpoint, breathing on Mars is going to be very expensive compared to breathing on Earth, too expensive if energy is hard to get.

Even if abundant fossil fuels were discovered on Mars, without a ready source of oxygen available for burning them, they would be useless as an energy source. Wind power isn't usable because, although winds may blow, given the extremely low atmospheric pressure, they will not be able to turn large-scale electric power–generating turbines. Solar cells and the necessary battery banks required for nighttime power will be used, but at best they can only provide enough energy for a basic level of existence with very limited manufacturing capabilities. Our bets are on thorium-based nuclear power using LFTRs (see chapter 4 for details). If LFTRs are not developed, some other form of nuclear reactor almost certainly will be required.

Given the right energy supply, Martians could live in underground structures and tunnels indefinitely even without terraforming the planet. Ironically, with no terraforming, it actually will be easier to stay warm on Mars than if the planet is fully terraformed. As mentioned in previous chapters, with a near-vacuum atmospheric pressure, there is little heat loss into it from habitats or from space suits. The lifestyle would be more confining than one in which the whole planet were terraformed to Earth's standards, but not too much different from living in a high-rise building located in a large metropolitan city on Earth. Keep in mind that Martians also will have awesome 3-D virtual-reality systems for entertainment. So even if Mars is not terraformed, it can be settled.

Most of our predictions could still be valid even without terraforming. However, with no motivation from the prospects of terraforming, Martians would not be crashing comets or asteroids into their planet, and hence would likely not be viewed as the masters of modifying asteroid orbits. Could Martians still be seen as a useful part of Earth's asteroid defense system? Yes, Martians could help observe and record asteroid movements and, in some cases, if remediation were needed to protect Earth from a dangerous asteroid, the hardware for the remediation might be launched and controlled from Mars.

Without the promise of terraforming, colonization and the conflicts that it produces would develop more slowly. Colonists will not be as motivated to emigrate massively without having the delusion that they are going to be able to roam the planet freely, possibly on vast sections of it that they own. Nevertheless, there will still be conflict between those who want to study Mars and those who want to develop it. Earthling investors and sponsoring governments will still be demanding and overly controlling of their Martian investments. So, even with little to no terraforming, conflict leading to rebellion eventually will build to the trigger point that causes Martians to demand their independence, although it may take longer to reach that point.

Will the demand for Martian independence automatically lead to an invasion by Earthlings? No, but it is a possibility and, let's face it, for us to discuss what a war with Martians would be like, we had to assume that the demand for Martian independence would not be accepted peacefully.

In many places we have attempted to discuss alternative scenarios, but for the sake of simplicity, we sometimes have chosen to ignore all but one. For the most part, we have assumed that the exploration and colonization of Mars will be driven largely by the United States. But it could be driven by China, or Europe, or Russia, or India. There could be multiple nations establishing settlements, as happened when the Europeans colonized the Americas, or maybe an international consortium. Even if largely driven by the United States, Martian colonization probably will be given a global-effort appearance with at least some international funding, similar to what was done with America's International Space Station.

Since the Mars colonization events discussed in this work have not yet become history, it's possible that we could be completely wrong about any or all of them. There is still value in contemplating them. The value is similar to that gained by a serious war game: It identifies the options and increases the ability to respond correctly to future events and avoid problems—even if those future events don't perfectly match the hypothetical ones studied. And at least some of the hypothetical events likely *will* happen.

It's risky to predict the future (that is, if one hopes to be right), and no sane people would do so in print, so here goes. Our predictions:

1. Mars will be colonized and become an important part of a space-mining effort, both on asteroids and on Mars itself, even if it is never fully terraformed.
2. Asteroid and/or Mars mining operations will play a significant role in giving Earth access to all the naturally occurring elements of the periodic table without all the pollution problems and shortage risks associated with many of the critically important elements.

3. Mars eventually will supplant Earth as the solar system's primary spacefaring planet.
4. A meteor with a kinetic energy in the range of 1 to 20 kilotons of TNT will crash into a populated area of the globe and cause mass casualties before Mars is fully terraformed. It will create hysteria to do something to prevent it from happening again. It is this smaller category of meteors, not the planet-killer category, that poses the greatest probability of killing a significant number of people.
5. Mars—and probably in the long term, Earth—will be powered primarily by nuclear energy using thorium in some form of LFTR.
6. Hydrogen-based fusion energy production is the energy wild card on Earth. If perfected, it will win as the primary energy source on Earth but probably not on Mars due to a lack of available hydrogen; even with terraforming, Mars will not be a wet planet for decades or, more likely, centuries.
7. Martian colonies will become independent of Earth both politically and with respect to basic needs.
8. Martians will demonstrate the ability to crash asteroids and comets into the Martian surface as a means of warming the planet and of bringing water or other desirable gases to the planet.

We have estimated that the independence of Mars is at least 250 years in the future. Historically, the most consistent error made by prognosticators, including science fiction writers, is to completely miss their time estimates, usually by underestimating them. For example, the 1968 movie *2001: A Space Odyssey* depicted a massive space station with artificial gravity in low Earth orbit, a well-established Moon base, and a gigantic manned spacecraft traveling to Saturn, all of which seem almost laughably advanced compared to what actually exists today, well after 2001. So, even in the best case (250 years to Martian independence), we won't be around to be embarrassed if it doesn't happen.

On a closing note, we would like briefly to address the possibility of an outer space–type war caused by an invasion from extraterrestrials (ETs). They would come from a totally different star system and need to travel light-years to get here. To do so, they would need to have technology that would make ours look prehistoric by comparison. Hopefully, they would be altruistic and come offering a cure for cancer, a complete set of plans for building a hydrogen-fueled fusion-driven power plant, and a cheap-as-dirt, light-as-air battery for powering our electric cars. Or they could be marauders intent on annihilating us and taking over our planet. In the latter case, we're going to be in big trouble, but it may not be hopeless.

Given the enormous amount of energy required to make the trip, the invaders are probably not going to arrive in a mother ship the size of a small moon containing numerous 15-mile diameter saucerlike spacecraft that will fly out of the ship to Earth and hover over major cities, waiting for the signal to make a coordinated attack (as depicted in the 1996 movie *Independence Day*).

ETs invading Earth will face many of the same difficulties encountered by Earthlings invading Mars. The ETs simply could fry Earth and wipe out all higher forms of life on it, but then the planet would essentially be wrecked. A more refined attack that focuses on annihilating or enslaving humanity will take far more time and resources—not an easy option given the length of the supply lines.

To pull off the more refined attack in the most efficient manner, the ETs would need a staging area to build up their forces, similar to the way England was used as a staging area to build up the D-Day invasion force in World War II. Possibly the best strategy might be to send a small preliminary unit with the genetic material and equipment needed to grow and eventually equip an invasion force by using local resources. This would take some time, but then highly advanced ETs might have highly advanced patience.

Options for locating the buildup would be limited. It's unlikely that it could be done on Earth without attracting unwanted attention and possibly an unwanted attack. In fact, humans could make ET life very hazardous if the ETs were located on Earth. The situation would be similar to the dangers from local inhabitants faced by early European colonists when they first arrived in America. Furthermore, even though Native Americans did not fully absorb European culture or technology, several Native American tribes proved to be very adroit at adopting European firearms and horses. So, given a chance to steal or otherwise acquire some ET technology, humans could possibly kick the ETs off the planet. If our assumption is that Mars could be inhabited by humans who would be capable of building an independent, self-sustaining society, then surely ETs with far superior technology could do the same thing. While the ETs amassing military might on Mars might be detected, realistically, what would humanity be able to do about it?

In closing, we have to admit that ETs using Mars for a staging area to attack Earth is not likely. For that matter, a war between Earthlings and Martians of any type is unlikely. If humanity is to benefit from colonizing Mars, however, we need to think ahead about the issues involved in doing so. Just leaving Mars alone as a barren planet in perpetuity does not seem like a good idea.

NOTES

1. Leah Crane, “Terraforming Mars Might Be Impossible Due to a Lack of Carbon Dioxide,” accessed June 6, 2019, https://www.newscientist.com/article/2175414-terraforming-mars-might-be-impossible-due-to-a-lack-of-carbon-dioxide/.

2. Stephen J. Bodio, “They Had Feathers: Is the World Ready to See Dinosaurs as They Really Were?,” accessed June 6, 2019, https://www.allaboutbirds.org/they-had-feathers-is-the-world-ready-to-see-dinosaurs-as-they-really-were-2/.

3. Clyde Haberman, “The Unrealized Horrors of Population Explosion,” accessed June 6, 2019, https://www.nytimes.com/2015/06/01/us/the-unrealized-horrors-of-population-explosion.html.

Bibliography

Abraham, David S. *The Elements of Power* (New Haven, CT: Yale University Press, 2015).

Agawa, Hiroyuki. *The Reluctant Admiral: Yamamoto and the Imperial Navy* (New York: Kodansha America, 1979).

Aldrin, Buzz, "Aldrin Mars Cycler," accessed June 5, https://buzzaldrin.com/space-vision/rocket_science/aldrin-mars-cycler/.

American Battlefield Trust, "American Revolution — FAQs," accessed June 5, 2019, https://www.civilwar.org/learn/articles/american-revolution-faqs.

Atkinson, Nancy, "What Are Asteroids Made of?," accessed June 5, 2019, https://www.universetoday.com/37425/what-are-asteroids-made-of/.

Atomic Heritage Foundation, "Espionage," accessed June 6, 2019, https://www.atomicheritage.org/history/espionage.

Biddle, Stephen. *Military Power: Explaining Victory and Defeat in Modern Battle* (Princeton, NJ: Princeton University Press, 2004).

Bluck, John, "Antarctic/Alaska-like Wind Turbines Could Be Used on Mars," NASA Ames Research Center, accessed June 5, 2019, https://www.nasa.gov/centers/ames/news/releases/2001/01_72AR.html.

Bodio, Stephen J., "They Had Feathers: Is the World Ready to See Dinosaurs as They Really Were?," accessed June 6, 2019, https://www.allaboutbirds.org/they-had-feathers-is-the-world-ready-to-see-dinosaurs-as-they-really-were-2/.

Boot, Max. *Invisible Armies: An Epic History of Guerrilla Warfare from Ancient Times to the Present* (New York: W. W. Norton, 2013).

Boulais, Richard, "How Long Does It Take for a Radio Signal to Go from Earth to Mars," accessed June 5, 2019, http://www.physlink.com/Education/AskExperts/ae381.cfm.

Bournias Varotsis, Alkaios, "Introduction to 3D Metal Printing," accessed June 5, 2019, https://www.3dhubs.com/knowledge-base/introduction-metal-3d-printing.

Bradford, Jason, "One Acre Feeds a Person," accessed June 5, 2019, http://www.farmlandlp.com/2012/01/one-acre-feeds-a-person/.

Brodwin, Erin, "NASA Sent Scott Kelly to Space for a Year, and 7% of His Genes Are Now Expressed Differently Than Those of His Identical Twin Mark," accessed June 5, 2019, https://www.businessinsider.com/nasa-twin-study-new-results-mark-scott-kelly-2017-10.

Burnham, James. *Suicide of the West: An Essay on the Meaning and Destiny of Liberalism* (New York: John Day, 1964).

Bush, Vannevar, et al., "Camouflage of Sea-Search Aircraft: Visibility Studies and Some Applications in the Field of Camouflage," Office of Scientific Research and Development, National Defense Research Committee, 1946., accessed June 5, 2019, https://apps.dtic.mil/dtic/tr/fulltext/u2/221102.pdf.

Buzan, Barry, Ole Weaver, and Jaap de Wilde. *Security: A New Framework for Analysis* (Boulder, CO: Lynne Rienner Publishers, 1997).

Carpenter, Charli. "Rethinking the Political Science/Fiction Nexus: The Campaign to Stop Killer Robots and Global Policy Making," *Perspectives on Politics* (Winter 2015).

Carson, Austin. "Facing Off and Saving Face: Covert Intervention and Escalation Management in the Korean War," *International Organization* 70, no. 1 (December 2016): 103–31.

Caspermeyer, Joe, "Space Flight Shown to Alter Ability of Bacteria to Cause Disease," accessed June 5, https://biodesign.asu.edu/news/space-flight-shown-alter-ability-bacteria-cause-disease.

Clinton, Henry. *The American Rebellion: Sir Henry Clinton's Narrative of His Campaigns, 1775–1782*, edited by William B. Willcox (New Haven, CT: Yale University Press, 1954).

CNN, "U.S. Nuclear Plant Had Partial Meltdown Years before Three Mile Island," accessed June 5, 2019, http://news.blogs.cnn.com/2011/03/29/u-s-nuclear-plant-had-partial-meltdown-years-before-three-mile-island/.

Collins, Graham P. "Claude E. Shannon: Founder of Information Theory," *Scientific American*, accessed June 5, 2019, https://www.scientificamerican.com/article/claude-e-shannon-founder/.

Conte Ronald L., Jr., "Hunter-Gatherers No More," accessed June 5, 2019, https://hungermath.wordpress.com/2013/11/05/hunter-gatherers-no-more/.

Coogan, Tim Pat, and George Morrison, *The Irish Civil War* (Boulder, CO: Roberts Rinehart Publishers, 1998).

Crane, Leah, "Terraforming Mars Might Be Impossible Due to a Lack of Carbon Dioxide," accessed June 6, 2019, https://www.newscientist.com/article/2175414-terraforming-mars-might-be-impossible-due-to-a-lack-of-carbon-dioxide/.

Crichton, Michael. *The Andromeda Strain* (New York: Vintage Books, 2017).

Daughan, George C. *Lexington and Concord: The Battle Heard Round the World* (New York: W. W. Norton, 2018).

David, Leonard, "Wind-Powered Mars Landers Could Really Work," accessed June 5, 2019, https://www.space.com/41023-mars-wind-power-landers-experiment.html.

D'Este, Carlo. *Fatal Decision: Anzio and the Battle for Rome* (New York: HarperCollins, 1991).

de Selding, Peter B. "SpaceX's New Price Chart Illustrates Performance Cost of Reusability," accessed June 5, 2019, https://spacenews.com/spacexs-new-price-chart-illustrates-performance-cost-of-reusability/.

Deneen, Patrick J. *Why Liberalism Failed* (New Haven, CT: Yale University Press, 2018).

Diamond, Jared. *Guns, Germs, and Steel: The Fates of Human Societies* (New York: W. W. Norton & Company, 1997).

Dorminey, Bruce, "Why Geothermal Energy Will Be Key To Mars Colonization," accessed June 5, 2019, https://www.forbes.com/sites/brucedorminey/2016/09/30/why-geothermal-energy-will-be-key-to-mars-colonization/#7eed970d4b25.

Dower, John W. *War Without Mercy: Race and Power in the Pacific War* (New York: Pantheon Books, 1986).

"Electrical Power Supply for Hostile Environments," Idaho National Laboratory, 2019, http://polarpower.org/PTC/2013_pdf/PTC_2013_Howe.pdf.

Emmanuelli, Matteo, "The Apollo 1 Fire," accessed June 5, 2019, http://www.spacesafetymagazine.com/space-disasters/apollo-1-fire/.

Environmental Protection Agency, "Understanding Global Warming Potentials," accessed June 5, 2019, https://www.epa.gov/ghgemissions/understanding-global-warming-potentials.

Farwell, Byron. *The Great Boer War* (Barnsley, South Yorkshire, Pen and Sword Military, 2009).

Fearon, James D., "Rationalist Explanations for War," *International Organization* 49. no. 3 (Summer 1995): 379–414.

Ferguson, Niall. *Civilization: The West and The Rest* (New York: Penguin Books, 2011).

Foer, Joshua, "Inside the Deep Caves Carved by Lava," accessed June 5, 2019, https://www.nationalgeographic.com/magazine/2017/06/lava-tubes-hawaii-caves/.

Forsythe, Paul, and Wolfgang A. Kunze, "Voices from Within: Gut Microbes and the CNS," accessed June 5, 2019, http://www.indiana.edu/~abcwest/pmwiki/CAFE/Voices%20from%20within-%20gut%20microbes%20and%20the%20CNS.pdf.

Freedberg Sydney J., Jr., "Lockheed, Army to Test Exoskeleton in December," accessed June 5, 2019, https://breakingdefense.com/2018/05/lockheed-army-to-test-exoskeleton-in-december/.

Fuller, John G. *We Almost Lost Detroit* (New York: Reader's Digest Press, 1975).

Gaget, Lucie, "3D Bioprinting: What Can We Achieve Today with a 3D Bioprinter?," accessed June 5, 2019, https://www.sculpteo.com/blog/2018/02/21/3d-bioprinting-what-can-we-achieve-today-with-a-3d-bioprinter.

Galula, David. *Counterinsurgency Warfare: Theory and Practice* (New York: Fredrick A. Praeger, 1964).

Garfield, Leanna, "This Robot Can 3D-print and Bake a Pizza in Six Minutes," accessed June 5, 2019, https://www.businessinsider.com/beehex-pizza-3d-printer-2017-3.

Gasperini, Luca, Enrico Bonatti, and Giuseppe Longo, "The Tunguska Mystery 100 Years Later," *Scientific American*, June 2008.

Gates, Bill, "Bill Gates Talks about Energy," Wall Street Journal YouTube, accessed June 5, 2019, https://www.youtube.com/watch?v=IsRlN1oDm60.

General Dynamics, "120mm KE-W A1® APFSDS-T," accessed June 5, 2019, https://www.gd-ots.com/munitions/large-caliber-ammunition/120mm-kew-a1/.

Glen, Alex, "Cost of Raising a Child Tops $260,000—Just for Basics," accessed June 5, 2019, https://www.nerdwallet.com/blog/insurance/cost-of-raising-a-child/.

Goldschein, Eric, "The Incredible Story of How De Beers Created and Lost the Most Powerful Monopoly Ever," accessed June 5, 2019, https://www.businessinsider.com/history-of-de-beers-2011-12.

Gordon, Peter, and Juan José Morales, "When the Dollar Spoke Spanish," accessed June 5, 2019, https://www.asiaglobalonline.hku.hk/when-the-dollar-spoke-spanish/.

Gorkin, Robert, and Susan Dodds, "The Ultimate Iron Chef: When 3-D Printers Invade the Kitchen," accessed June 5, 2019, https://phys.org/news/2013-10-ultimate-iron-chef-d-printers.html.

Griffith, Paddy. *Battle Tactics of the Civil War* (New Haven, CT: Yale University Press, 1989).

Grossman, Dave. *On Killing: The Psychological Cost of Learning to Kill in War and Society* (New York: Back Bay Books, 2009).

Gurr, Ted Robert. *Why Men Rebel: Fortieth Anniversary Edition* (Boulder, CO: Paradigm Publishers, 2010).

Gyeltshen, Jamba, and Amanda Hodges, "Japanese Beetles," accessed June 5, 2019, http://entnemdept.ufl.edu/creatures/orn/beetles/japanese_beetle.htm.

Haas, Lawrence. *Harry and Arthur: Truman, Vandenberg, and the Partnership that Created the Free World* (Lincoln, NE: Potomac Books, 2016).

Haberman, Clyde, "The Unrealized Horrors of Population Explosion," accessed June 6, 2019, https://www.nytimes.com/2015/06/01/us/the-unrealized-horrors-of-population-explosion.html.

Hargraves, Robert, "Thorium Energy Cheaper Than Coal," Amazon Digital Services LLC, Kindle Edition: loc. 3023.

Hargraves, Robert, and Ralph Moir, "Liquid Fuel Nuclear Reactors," Forum on Physics and Society, APS Physics, accessed June 5, 2019, https://www.aps.org/units/fps/newsletters/201101/hargraves.cfm.

Harkins, Gina, "The Marines Just 3D-Printed an Entire Bridge in California," accessed June 5, 2019, https://www.military.com/dodbuzz/2019/01/31/marines-just-3d-printed-entire-bridge-california.html.

Harris, William, "How Lunar Liquid Mirror Telescopes Work," accessed June 5, 2019, https://science.howstuffworks.com/liquid-mirror-telescope1.htm.

Healy, Mark. *Cannae 216 BC: Hannibal Smashes Rome's Army* (Oxford: Osprey Publishing, 1994).

Heron, Gil Scott, "We Almost Lost Detroit," accessed June 5, 2019, https://www.youtube.com/watch?v=yotCw66_G1g.

Hess, Earl J. *The Rifle Musket in Civil War Combat: Myths and Reality* (Lawrence: The University Press of Kansas, 2008).

Heward, Anita, "Lava Tubes as Hidden Sites for Future Human Habitats on the Moon and Mars," accessed June 5, 2019, https://phys.org/news/2017-09-lava-tubes-hidden-sites-future.html.

Homer-Dixon, Thomas. *The Upside of Down: Catastrophe, Creativity, and the Renewal of Civilization* (Washington, DC: Island Press, 2006).

Hornfischer, James D. *Last Stand of the Tin Can Sailors: The Extraordinary World War II Story of the U.S. Navy's Finest Hour* (New York: Bantam, 2004).

———. *Neptune's Inferno: The U.S. Navy at Guadalcanal* (New York: Bantam, 2011).

Howe, Steven D., Robert C. O'Brien, Troy M. Howe, Carl Stoots, "Compact, Low Specific-Mass Electrical Power Supply for Hostile Environments," Idaho National Laboratory, accessed Oct. 2019, http://polar.sri.com/polarpower.org/PTC/2013_pdf/PTC_2013_Howe.pdf.

Ikle, Fred Charles. *Every War Must End: Second Revised Edition* (New York: Columbia University Press, 2005).

Jackson, Patrick Thaddeus. *Civilizing the Enemy: German Reconstruction and the Invention of the West* (Ann Arbor: University of Michigan Press, 2006).

Jennings, Ed, "Crosley's Secret War Effort-The Proximity Fuze," accessed June 5, 2019, http://www.navweaps.com/index_tech/tech-075.php.

Jervis, Robert, "Cooperation under the Security Dilemma," *World Politics* 30, no.2 (January 1978): 167–74.

———. *Perception and Misperception in International Politics* (Princeton, NJ: Princeton University Press, 1976).

Joseph, Rhawn Gabriel, "Space Fungi Are Attacking the Space Stations," accessed June 5, 2019, http://cosmology.com/SpaceFungi.html.

Keeley, Lawrence H. *War before Civilization: The Myth of the Peaceful Savage* (Oxford: Oxford University Press, 1997).

Kelso, William N. *Jamestown the Truth Revealed* (Charlottesville: University of Virginia Press, 2017).

Kestenbaum, David, "Spaceflight Is Getting Cheaper. But It's Still Not Cheap Enough," accessed June 5, 2019, https://www.npr.org/sections/money/2011/07/21/138166072/spaceflight-is-getting-cheaper-but-its-still-not-cheap-enough.

Kingery, Ken, "Beyond Materials: From Invisibility Cloaks to Satellite Communications," accessed June 5, 2019, https://stories.duke.edu/beyond-materials-from-invisibility-cloaks-to-satellite-communications.

Knighton, Andrew, "The Battle of 73 Eastings—The Mother of All Battles?," accessed June 6, 2019, https://www.warhistoryonline.com/featured/battle-73-easting.html/2.

Krasner, Stephen D. *Sovereignty; Organized Hypocrisy* (Princeton, NJ: Princeton University Press, 1999).

Krotoski, Aleks. *Untangling the Web: What the Internet Is Doing to You.* (London: Guardian Books, 2013).

Küppers, Michael, Laurence O'Rourke, Dominique Bockelée-Morvan, Vladimir Zakharov, Seungwon Lee, Paul von Allmen, Benoît Carry, David Teyssier, Anthony Marston, Thomas Müller, Jacques Crovisier, M. Antonietta Barucci, and Raphael Moreno, "Localized Sources of Water Vapour on the Dwarf Planet Ceres," *Nature*, Vol. 505 (January 23, 2014): 525–27.

Landis, Geoffrey A., "Human Exposure to Vacuum," accessed June 5, 2019, http://www.geoffreylandis.com/vacuum.html.

Lant, Karla, "NASA Is Fast-Tracking Plans to Explore a Metal Asteroid Worth $10,000 Quadrillion," accessed June 5, 2019, https://futurism.com/nasa-fast-tracking-plans-explore-metal-asteroid-worth-10000-quadrillion/.

Lasswell, Harrold Dwight. *Politics: Who Gets What, When, How* (New York: McGraw Hill, 1936).

Lawrence, T.E. *Seven Pillars of Wisdom* (Ware, Hertfordshire: Bibliophile Books, 1997).

Legro, Jeffrey W. *Cooperation Under Fire: Anglo-German Restraint During World War II* (Ithaca: Cornell University Press, 1995).

"Lithium Ion Battery," Clean Energy Institute, University of Washington, 2019, https://www.cei.washington.edu/education/science-of-solar/battery-technology/.

Lo, Chris, "The False Monopoly: China and the Rare Earths Trade," accessed June 5, 2019, https://www.mining-technology.com/features/featurethe-false-monopoly-china-and-the-rare-earths-trade-4646712/.

Lukas, John. *Outgrowing Democracy: A History of the United States in the Twentieth Century* (New York: Doubleday, 1984).

MacMillan, Margaret. *Paris 1919: Six Months that Changed the World* (New York: Random House, 2002).

Mahaffey, James. *Atomic Awakening* (New York: Pegasus Books, 2010), Kindle Edition.

Malaska, Mike, "Earth's Toughest Life Could Survive on Mars, Planetary Society," accessed June 5, 2019, http://www.planetary.org/blogs/guest-blogs/20120515-earth-life-survive-mars.html.

Malthus, Thomas. *An Essay on the Principle of Population* (London: J. Johnson, 1798).

Mambra, Shamseer, "USS Nimitz: One of the Biggest War Ships in the World," accessed June 5, 2019, https://www.marineinsight.com/types-of-ships/uss-nimitz-one-of-the-biggest-war-ships-in-the-world/.

Marshall, S.L.A. *Men against Fire: The Problem of Battle Command* (Norman: University of Oklahoma Press, 2000).

Matias, Elizabeth, Bharat Rao, "3D Printing: On Its Historical Evolution and the Implications for Business," 2015 Proceedings of PICMET '15: Management of the Technology Age, http://faculty.poly.edu/~brao/3dppicmet.pdf.

Menker, Sarah, "Like Day and Nitrogen: War, Peace, and the Dawn of Fertilizers," accessed June 5, 2019, https://www.linkedin.com/pulse/like-day-nitrogen-war-peace-dawn-fertilizers-sara-menker/?articleId=8845390223113810724.

Meuller, John. *Overblown: How Politicians and the Terrorism Industry Inflate National Security Threats, and Why We Believe Them* (New York: Free Press, 2006).

Miksha, Ron, "The Man Who Made Killer Bees," accessed June 5, 2019, https://badbeekeepingblog.com/2016/09/09/the-man-who-made-killer-bees/.

Miller, Seumas. *Dual Use Science and Technology, Ethics and Weapons of Mass Destruction* (Cham, Switzerland: Springer International Publishers, 2018).

Mizokami, Kyle, "Become a Super Sniper: DARPA Is Turning .50 Caliber Bullets into Guided Rounds," accessed June 6, 2019, https://nationalinterest.org/blog/buzz/become-super-sniper-darpa-turning-50-caliber-bullets-guided-rounds-27101.

Mori, Masahiro, "The Uncanny Valley: The Original Essay," accessed June 5, 2019 https://spectrum.ieee.org/automaton/robotics/humanoids/the-uncanny-valley.

Mosher, Dave, "A Cold War Technology Designed to Make Jets Fly for Days Might Solve Earth's Looming Energy Crisis," accessed June 5, 2019, https://www.businessinsider.com/aircraft-nuclear-propulsion-molten-salt-reactor-2016-12.

Musk, Elon, "Elon Musk Reveals His Plan for Colonizing Mars," accessed June 5, 2019, https://www.youtube.com/watch?v=W9olSzNOh8s.

Nagl, John A. *Learning to Eat Soup with a Knife: Counterinsurgency Lessons from Malaya and Vietnam* (Chicago: University of Chicago Press, 2005).

NASA, "Advanced Space Transportation Program: Paving the Highway to Space," accessed June 5, 2019, https://www.nasa.gov/centers/marshall/news/background/facts/astp.html.

———, "Ion Propulsion," accessed June 5, 2019, https://solarsystem.nasa.gov/missions/dawn/technology/ion-propulsion/.

———, "Liquid Mirror Telescopes on the Moon," accessed June 5, 2019, https://science.nasa.gov/science-news/science-at-nasa/2008/09oct_liquidmirror.

National Center for Education Statistics, accessed June 5, 2019, https://nces.ed.gov/fastfacts/display.asp?id=66.

———, accessed June 5, 2019, https://nces.ed.gov/fastfacts/display.asp?id=76.

National Park Service—Historic Jamestowne, "The First Residents of Jamestown," accessed June 5, 2019, https://www.nps.gov/jame/learn/historyculture/the-first-residents-of-jamestown.htm.

Navweaps.com, "Electromagnetic Rail Gun Proposal," accessed June 5, 2019, http://www.navweaps.com/Weapons/WNUS_Rail_Gun.php.

Nield, David, "Our Smartphone Addiction Is Costing the Earth," accessed June 5, 2019, https://www.techradar.com/news/phone-and-communications/mobile-phones/our-smartphone-addiction-is-costing-the-earth-1299378.

Nixon, Richard, "Nixon Phone Call—Nixon Speech on Jobs in California," accessed June 5, 2019, https://www.youtube.com/watch?v=Mj5gFB5kTo4.

———, "Remarks at the Atomic Energy Commission's Hanford Works near Richland, Washington," accessed June 5, 2019, http://www.presidency.ucsb.edu/ws/index.php?pid=3161.

NOAA, "Official U.S. Government Information about the Global Positioning System (GPS) and Related Topics," accessed June 6, 2019, www.GPS.gov.

Nuclear Security.com, "Nukemap V. 2.61," accessed June 5, 2019, http://www.nuclearsecrecy.com/nukemap/.

Office of Energy Efficiency and Renewable Energy, 2019, https://www.energy.gov/eere/fuelcells/hydrogen-storage-basics-0.

Onion, Amanda, "Was Germany Doomed in World War I by the Schlieffen Plan?," accessed June 5, 2019, https://www.history.com/news/was-germany-doomed-in-world-war-i-by-the-schlieffen-plan.

O'Shaughnessy, Andrew Jackson. *The Men Who Lost America: British Leadership, the American Revolution, and the Fate of the Empire* (New Haven, CT: Yale University Press, 2013).

Pomeroy, Ross, "Why Not Nuclear-Powered Aircraft?," accessed June 5, 2019, http://www.realclearscience.com/blog/2014/07/why_not_nuclear-powered_aircraft.html.

Powell, Andrea, "How to 3D Print an Entire House in a Single Day," accessed June 5, 2019, https://www.wired.com/story/icon-house-3d-printer/.

Pushkar, Robert G., "Comet's Tale," accessed June 5, 2019, https://www.smithsonianmag.com/history/comets-tale-63573615/.

Ramirez, Ainissa, "Where to Find Rare Earth Elements," Nova Next, accessed June 5, 2019, http://www.pbs.org/wgbh/nova/next/physics/rare-earth-elements-in-cell-phones/.

Rask, Jon, M.S., Wenonah Vercoutere, Ph.D., Barbara J. Navarro, and Al Krause, M.S., "An Interdisciplinary Guide on Radiation and Human Space Flight," NASA, accessed June 5, 2019, https://www.nasa.gov/pdf/284273main_Radiation_HS_Mod1.pdf.

Redd, Nola Taylor, "OSIRIS-Rex: Bringing Home Pieces of an Asteroid," accessed June 5, 2019, https://www.space.com/33776-osiris-rex.html.

Reedy, Christianna, "NASA Wants to Collect Solar Power Directly From Space," accessed June 5, 2019, https://futurism.com/is-space-based-solar-power-the-answer-to-our-energy-problem-on-earth/.

Rejcek, Peter, "A Mars Survival Guide: Finding Food, Water, and Shelter on the Red Planet," accessed June 5, 2019, https://singularityhub.com/2017/05/28/a-mars-surival-guide-how-to-find-food-water-and-shelter-on-the-red-planet/#sm.00001tfrpjmixfcwzqz741lzt31f.

Rex Plastics, "How Much Do Plastic Injection Molds Cost?," accessed June 5, 2019, https://rexplastics.com/plastic-injection-molds/how-much-do-plastic-injection-molds-costww.

Richardot, Amandine, "Can You 3D Print a 3D Printer?," accessed June 5, 2019, https://www.sculpteo.com/blog/2017/10/24/3d-print-3d-printer/?&&&.

Riddle, Lincoln, "The Cu Chi Tunnels: A Dangerous Underground Warzone in the Vietnam War," accessed June 5, 2019, https://www.warhistoryonline.com/vietnam-war/cu-chi-tunnels-dangerous-underground-warzone-b.html.

Robinson, Anthony Frank Hardisty, and George Chaplin, "Technology Trends—Smart Dust and Sensor Networks," accessed June 6, 2019, https://www.e-education.psu.edu/geog583/node/77.

Roblin, Sebastien, "That Time the Allies Engineered a 'Cloaking Device' during World War II," accessed June 5, 2019, https://taskandpurpose.com/allies-cloaking-device-world-war-ii/.

Russell, Shahan, "The USS Johnston and Its Kamikaze Captain, Took on 4 Battleships, 8 Cruisers, and 11 Destroyers at Leyte Gulf," accessed June 5, 2019, https://www.warhistoryonline.com/world-war-ii/uss-johnston-kamikazecaptain.html.

Russia Times, "Meteorite Hits Russian Urals: Fireball Explosion Wreaks Havoc, Up to 1,200 Injured," accessed June 5, 2019, https://www.rt.com/news/meteorite-crash-urals-chelyabinsk-283/.

Sainato, Michael, "Stephen Hawking, Elon Musk, and Bill Gates Warn about Artificial Intelligence," accessed June 5, 2019, http://observer.com/2015/08/stephen-hawking-elon-musk-and-bill-gates-warn-about-artificial-intelligence/.

Sandle, Tim, "New Approach to Treating Depression with Brain Implants," accessed June 6, 2019, http://www.digitaljournal.com/tech-and-science/science/new-approach-to-treating-depression-with-brain-implants/article/538060.

Schelling, Thomas C. *Arms and Influence* (New Haven, CT: Yale University Press, 1966).

Schlachter, Fred, Has the Battery Bubble Burst?," APS News, August/September 2012, https://www.aps.org/publications/apsnews/201208/backpage.cfm.

Sciutto, Jim, and Jennifer Rizzo, "War in Space: Kamikazes, Kidnapper Satellites and Lasers," accessed June 6, 2019, https://www.cnn.com/2016/11/29/politics/space-war-lasers-satellites-russia-china/index.html.

Scoles, Sarah, "NASA Likely to Break Radiation Rules to Go to Mars," accessed June 5, 2019, http://www.pbs.org/wgbh/nova/next/space/nasa-mars-radiation-rule/.

Sharman, J. C. *Empires of the Weak: The Real Story of European Expansion and the Creation of the New World Order* (Princeton, NJ: Princeton University Press, 2019).

Sharp, Tim, "How Big Is Mars?," accessed June 5, 2019, https://www.space.com/16871-how-big-is-mars.html.

———, "What Is the Temperature of Mars?," accessed June 5, 2019, https://www.space.com/16907-what-is-the-temperature-of-mars.html.

Shaw, Stephen, "Posts Tagged 'C-type asteroids' Asteroid Mining," accessed June 5, 2019, http://www.astronomysource.com/tag/c-type-asteroids/.

Smartech Analysis, "Use of 3D Bioprinting in Drug Discovery and Cosmetics Testing Expected to Reach $500 Million by 2027," accessed June 6, 2019, https://www.smartechanalysis.com/blog/3d-bioprinting-drug-discovery/.

Snowden, Scott, "Virgin Galactic's Epic 1st Spaceflight Inspires Richard Branson's Message to Grandkids," accessed June 5, 2019, https://www.space.com/42809-virgin-galactic-first-spaceflight-richard-branson-video.html.

Snyder, Jack. *Myths of Empire: Domestic Politics and International Ambition* (Ithaca, NY: Cornell University Press, 1991).

Song, Hilary Huaici, "Nuclear Weapons 101: Back to the Basics," Columbia University K1 Project Center for Nuclear Studies, accessed June 5, 2019, https://k1project.columbia.edu/news/nuclear-weapons-101-back-basics.

SpaceX, "Making Life Multiplanetary," accessed June 5, 2019, https://www.spacex.com/mars.

Springer, Mike, "Learn How Richard Feynman Cracked the Safes with Atomic Secrets at Los Alamos," accessed June 6, 2019, http://www.openculture.com/2013/04/learn_how_richard_feynman_cracked_the_safes_with_atomic_secrets_at_los_alamos.html.

Sputnik News, "DARPA's New Brain Chip Enables Telepathic Control of Drone Swarms," accessed June 6, 2019, https://sputniknews.com/military/201809071067847857-Brain-Chip-Fly-Drone-Swarm-Telepathically/.

Stanlis, Peter, ed. *Edmund Burke: Selected Writings and Speeches* (New York: Routledge, 2017).

Stromberg, Joseph, "Starving Settlers in Jamestown Colony Resorted to Cannibalism," accessed June 5, 2019, http://www.smithsonianmag.com/history/starving-settlers-in-jamestown-colony-resorted-to-cannibalism-46000815/.

Summers, Frank, "Angular Resolution and What Hubble Can't See, Hubble's Universe," accessed June 5, 2019, http://hubblesite.org/explore_astronomy/hubbles_universe_unfiltered/blogs/angular-resolution-and-what-hubble-cant-see.

Svoboda, Elizabeth, "The Truth About Thorium and Nuclear Power," accessed June 5, 2019, http://www.popularmechanics.com/science/energy/a6162/the-truth-about-thorium-and-nuclear-power/.

Tannenwald, Nina. *The Nuclear Taboo: The United States and the Non-Use of Nuclear Weapons since 1945* (Cambridge: Cambridge University Press, 2007).

Teitel, Amy Shira, "Why Did NASA Still Use Pure Oxygen after the Apollo 1 Fire?," *Popular Science*, accessed June 5, 2019, https://www.popsci.com/why-did-nasa-still-use-pure-oxygen-after-apollo-1-fire.

Thompson, Clive, "When Robots Take All of Our Jobs, Remember the Luddites," accessed June 5, 2019, https://www.smithsonianmag.com/innovation/when-robots-take-jobs-remember-luddites-180961423/.

Tilley, David, "Platinum Group Metals," accessed June 5, 2019, https://www.geologyforinvestors.com/platinum-group-metals/.

Tse-tung, Mao. *On Guerilla Warfare*, translated Samuel B. Griffith II (Champaign: University of Illinois Press, 2000).

Tzu, Sun. *The Art of War: Everyman's Edition*, Trans. Peter Harris (New York: Alfred A. Knopf, 2018).

United Nations Office for Outer Space Affairs, "Near Earth Objects and Planetary Defense," accessed June 5, 2019, http://www.unoosa.org/documents/pdf/smpag/st_space_073E.pdf.

———, "Treaty on Principles Governing the Activities of States in the Exploration and Use of Outer Space, including the Moon and Other Celestial Bodies," accessed June 5, 2019, http://www.unoosa.org/oosa/en/ourwork/spacelaw/treaties/outerspacetreaty.html.

United States Nuclear Regulatory Commission, accessed June 5, 2019, https://www.nrc.gov/materials/fuel-cycle-fac/ur-enrichment.html.

Vijayan, Jaikumar, "Infected USB Drive Blamed for '08 Military Cyber Breach," accessed June 6, 2019, https://www.computerworld.com/article/2514879/security0/infected-usb-drive-blamed-for--08-military-cyber-breach.html.

Wallace, Arminta, "Mary Mallon, the Irish Woman Who Brought Typhoid to New York," accessed June 5, 2019, https://www.irishtimes.com/life-and-style/abroad/mary-mallon-the-irish-woman-who-brought-typhoid-to-new-york-1.3125437.

Wallentine, Cynthia, "How NASA's Planetary Protection Officer Keeps Our Germs from Contaminating Other Planets (& Vice Versa)," accessed June 5, 2019, https://invisiverse.wonderhowto.com/news/prime-directive-nasas-planetary-protection-officer-keeps-our-germs-from-contaminating-other-planets-vice-versa-0176749/.

Warfare History Network, "How the U.S. Military Went to War against Vietnam's Radar and Air Defenses," accessed June 5, 2019, https://nationalinterest.org/blog/the-buzz/how-the-us-military-went-war-against-vietnams-radar-air-25034.

Watkin, Hanna, "Researchers Developing Self-Replicating 3D Printers to Build Moon Bases," accessed June 5, 2019, https://all3dp.com/carleton-university-researchers-develop-self-replicating-3d-printers-help-build-moon-bases/.

Watson Institute for International and Public Affairs, "Costs of War," accessed June 5, 2019, https://watson.brown.edu/costsofwar/.

Weber, Max. *The Protestant Work Ethic and the Spirit of Capitalism: And Other Writings*, edited by Peter Baehr and Gordon C. Welle (New York: Penguin, 2002).

Wilkinson, Stephan, "The Goldilocks Fighter: The F6F Hellcat," accessed June 6, 2019, http://www.historynet.com/goldilocks-fighter-f6f-hellcat.htm.

World Nuclear Association, "Nuclear Power Reactors," accessed June 5, 2019, http://www.world-nuclear.org/information-library/nuclear-fuel-cycle/nuclear-power-reactors/nuclear-power-reactors.aspx.

Worral, Simon, "How Racism, Arrogance, and Incompetence Led to Pearl Harbor," accessed June 5, 2019, https://news.nationalgeographic.com/2016/12/countdown-pearl-harbor-attack-twomey-anniversary/.

Worthington, Daryl, "Spanish Armada Sets Sail from Corunna," accessed June 5, 2019, https://www.newhistorian.com/spanish-armada-sets-sail-from-corunna/4360/.

Yonck, Richard, "The Dawn of Space Age Mining," accessed June 5, 2019, https://blogs.scientificamerican.com/guest-blog/the-dawn-of-the-space-mining-age/.

Zhang, Moran, "Russia Meteor 2013: Damage to Top $33 Million; Rescue, Cleanup Team Heads to Meteorite-Hit Urals," accessed June 5, 2019, https://www.ibtimes.com/russia-meteor-2013-damage-top-33-million-rescue-cleanup-team-heads-meteorite-hit-urals-1090104

Zhang, Ping, "Energy Density of Fats," *The Physics Fact Book, 2004,* https://hypertextbook.com/facts/2004/PingZhang.shtml.

Zubrin, Robert. *The Case for Mars* (New York: Free Press, 1996).

Index

About the Authors

T. K. Rogers is an award-winning STEM educator who has a bachelor's degree in mechanical engineering from Arizona State University and an MBA from Clemson University. He is a U.S. patentholder who worked as an engineer in the chemical and plastics industries for 18 years before becoming a high school STEM educator, a career he pursued for over 21 years, mostly as part of the International Baccalaureate program in Greenville, South Carolina. As part of this program, during a typical year, he taught calculus-based Advanced Placement Physics, Advanced Placement Statistics, and Advanced Placement Computer Science.

In 1997 Rogers sought to extend beyond the classroom and founded Insultingly Stupid Movie Physics, a website devoted to revealing the physics nonsense in Hollywood movies, designed to both entertain and enlighten. In 2007 Source Books published his book with the same title. His other publications include the essay "Fellow Nerds: Let's Celebrate Nerdiness," first published in the December 11, 2000, edition of *Newsweek* magazine.

After retiring from classroom teaching Rogers now resides with his wife and three dogs in a passive solar house of his own design located on his urban homestead near Greer, SC. Similar to one of the characters in the movie "Ghost Busters" Rogers collects fungus, in this case shiitake, reishi, and oyster mushrooms that he deliberately grows as part of his homesteading efforts.

J. Furman Daniel, III is an assistant professor in the College of Security and Intelligence at Embry-Riddle Aeronautical University. He holds a BA *cum laude* in history and political science from the University of Chicago, and a PhD in government from Georgetown University.

Professor Daniel's research and teaching span a wide range of topics including military strategy, political thought, wargaming, and international relations theory. He is the editor of *21st Century Patton: Leadership Lessons for the Modern Era* (Naval Institute Press, 2016) and the author of the forthcoming book, *Patton: Battling with History* (University of Missouri Press). His work has appeared in the *Washington Post*, *Real Clear Defense*, *The Guardian Anti-Terrorism Journal*, *Small Wars Journal*, *Washington Quarterly*, *U.S. Naval Institute Proceedings*, *Polity*, *The Journal of International Political Theory*, *Orbis*, and *International Studies Quarterly*.

In addition to his scholarly activities, Professor Daniel serves as an officer in the United States Navy Reserves, is a rabid baseball fan, and is happily married to the love of his life, Christina Capacci-Daniel, PhD. He lives in Prescott, Arizona.

www.ingramcontent.com/pod-product-compliance
Lightning Source LLC
Chambersburg PA
CBHW051253200926
51612CB00001B/26
9781498587747